Transportation Systems and Service Policy

A Project-Based Introduction

Transportation Systems and Service Policy

A Project-Based Introduction

John G. Schoon
Professor Emeritus
Northeastern University

CHAPMAN & HALL

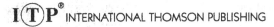

I(T)P® INTERNATIONAL THOMSON PUBLISHING

New York • Albany • Bonn • Boston • Cincinnati • Detroit • London • Madrid • Melbourne
Mexico City • Pacific Grove • Paris • San Francisco • Singapore • Tokyo • Toronto • Washington

Cover design: Curtis Tow Graphics

Printed in the United States of America

Chapman & Hall
115 Fifth Avenue
New York, NY 10003

Chapman & Hall
2-6 Boundary Row
London SE1 8HN
England

Thomas Nelson Australia
102 Dodds Street
South Melbourne, 3205
Victoria, Australia

Chapman & Hall GmbH
Postfach 100 263
D-69442 Weinheim
Germany

International Thomson Editores
Campos Eliseos 385, Piso 7
Col. Polanco
11560 Mexico D.F
Mexico

International Thomson Publishing–Japan
Hirakawacho-cho Kyowa Building, 3F
1-2-1 Hirakawacho-cho
Chiyoda-ku, 102 Tokyo
Japan

International Thomson Publishing Asia
221 Henderson Road #05-10
Henderson Building
Singapore 0315

1 2 3 4 5 6 7 8 9 10 XXX 01 00 99 98 97 96

Library of Congress Cataloging-in-Publication Data

Schoon, John G., 1937–
    Transportation systems and service policy : a project-based
introduction / by John G. Schoon
        p.   cm.
    Includes bibliographical references and index.
    ISBN 0-412-07481-8 (pbk. : alk. paper)
    1. Transportation engineering.   2. Transportation — Planning.
I. Title.
TA1145.S33   1996
388′ .068--dc20   95-48465
                                                          CIP

British Library Cataloguing in Publication Data available

To order this or any other Chapman & Hall book, please contact **International Thomson Publishing, 7625 Empire Drive, Florence, KY 41042.** Phone: (606) 525-6600 or 1-800-842-3636. Fax: (606) 525-7778. e-mail: order@chaphall.com.

For a complete listing of Chapman & Hall titles, send your request to **Chapman & Hall, Dept. BC, 115 Fifth Avenue, New York, NY 10003.**

Table of Contents

Preface

The many aspects of urban transportation planning and design demand a multi-faceted approach to ensure responsive, economical, and environmentally sensitive facilities that enhance mobility. Yet all too easily the complexity of the process can obscure the major elements. This book aims at assisting the analyst to provide decision makers with a range of solutions by illustrating how service policies regarding quality of service, fares, investment levels, and environmental impacts affect and are affected by each other.

This book, therefore, concentrates on the process of planning and design. It addresses the major elements of urban transportation planning, design, and impact estimation, and offers practice in undertaking typical projects. It focuses on the linkages and interaction with public policy regarding user service levels, and the resulting design and impacts. The process is illustrated by (1) outlining the individual transportation analysis and design techniques and their linkages, (2) describing the planning and design process, from population changes affecting demand and mobility needs to estimation of air pollution and energy use impacts that are instrumental in shaping public policy and strategic planning, (3) presenting examples of transportation design projects showing how service policy may affect the physical and operational design of multimodal, urban transportation systems, (4) enabling the readers to obtain practice in basic, applied transportation analysis, design, and impact estimation by defining the key service policy variables of projects for solution, and (5) familiarizing the reader with some of the major transportation planning and design procedures, as described in readily available documents.

The book complements the many extensive texts and government documents on transportation theory and analysis. It focuses on the *application* of generally accepted procedures to assist senior undergraduate students, graduate students, and practitioners who have a basic knowledge of transportation systems design, or who are studying the theory, analysis, and characteristics of urban transportation systems in greater detail, concurrently with this book. It is hoped that this will help provide a perspective on the total process, so essential to understanding and participating in effective program and project implementation, and also maximize effective project control and administration. Reviews and design modifications are integral parts of the creative transportation design process, providing immense challenges in combining technical excellence with established and evolving public policy. The material also reviews selected fundamental concepts that will enable those with backgrounds in economics, urban and regional planning, and related disciplines to become aware of the kinds and the scope of projects likely to be encountered in practice. The design content and projects for solution are also structured to provide a basis for design course requirements for undergraduate civil engineering degree accreditation. Following the introduction in Part 1, the book comprises two parts, both of which address essential features of urban transportation planning:

Part 2: Long range, multimodal transportation system planning and design

Part 3: Short-range transportation demand management and design.

The implementation period (from concept to implementation) of an actual transportation project may vary from less than a year for a transportation demand management (TDM) project to 20 or more years for an extensive corridor project. Usually, however, the project would be a part of a continuing program responding to emerging needs and issues, and would almost invariably be modified several times before implementation.

By using the preliminary or "sketch planning" approach, the essential features of the process are retained and the detailed computations are minimized to encourage a sense of the *total process* as a background for more detailed investigations. Approximations are noted and deemed consistent with sketch planning methods of evaluating the effects of different values of policy variables before specific projects are selected for detailed analysis and eventual implementation. More specifically, the objectives of the book are to:

- Specify selected transportation policy options and conduct the planning, design, and impact estimation of a typical urban transportation corridor project ranging from identification of demographic changes to evaluation of alternative designs (Part 2), and to explore the impacts of selected transportation system and demand management options (Part 3).

- Complement existing texts, manuals, guidelines, and other material related to transportation planning, design, and evaluation to provide an example project and a number of alternative scenarios embodying practical policy options. This encourages the development of creative, workable solutions and enables the mechanisms of change and systems management to be demonstrated and explored by the instructor, individual, or groups of students, using a common methodology but different input values.

- Present the analyses in a simple yet realistic fashion to enable computations to be completed by use of a pocket calculator as well as by microcomputer spreadsheets or computer programs prepared by the student.

- Avoid the use of techniques where assumptions and methods may not be readily discerned and understood.

- Enable the student or instructor to intervene at various points in the process in order to experiment by varying the policy variables, and to encourage debate, especially concerning the impact levels of various strategies and how these may relate to ongoing issues in transportation analysis, planning, and development.

- Use, whenever possible, generally accepted and readily available data and analysis techniques in the illustrative examples.

- Illustrate where current planning and design techniques may be deficient, and where improvements may be desirable.

The approach to describing the planning and design process is illustrated by the sequence and content of the chapters, as follows.

Introduction

Chapter 1 outlines the typical transportation planning processes and the basic relationships between long- and short-range planning. The importance of integrating technical and policy elements and the basis for establishing the latter are stressed. Major features of transportation corridors are outlined. The corridor level of planning is the one used to illustrate the planning and design process in the ensuing chapters.

Part 2: Long-Range Transportation Planning

Chapter 2 describes the techniques used in the traditional sequential transportation demand estimation process, along with examples presented in a format that can be used to analyze the projects presented later. This provides a basis for **Chapter 3,** where the facilities design features are shown to have a direct effect on the nature and extent of the impacts—particularly capital cost, energy use, and air pollution. This chapter first outlines methods of initially selecting appropriate mixes of modes, including automobile, bus, and rail transit, in accordance with defined levels of service. It then indicates the characteristics of the impacts associated with these modes and how the combinations may be refined to better respond to the imposed service levels and resulting impacts.

Chapter 4 provides an example for conducting an individual project. Illustrative physical and operational characteristics of a typical long-range intermodal corridor plan, to be completed at the sketch planning level of detail, are presented. The mechanisms for formulating the design for a given set of transportation service policy alternatives, conducting the demand analysis, preparing a preliminary physical design, and estimating key impacts of capital cost, energy use, air pollution, and passenger travel time are shown. *This provides the basic information and scope of work needed to investigate the impacts of a number of alternatives with specified policy variables ranging between automobile intensive to transit intensive, in response to national, state, and urban priorities. Each alternative project is intended to be prepared by one or two participants in a transportation planning and engineering course, and typifies the efforts of a design organization. The intermediate results and a range of final impacts for each project are provided in an* **instructor's manual** *to assist in monitoring progress and interpreting the final results.*

Part 3: Short-Range Transportation Systems Planning

Chapter 5 presents background information on High Occupancy Vehicle (HOV) projects, their role in the transportation system, and factors affecting their feasibility. Examples of estimating the demand for an HOV facility and the associated changes in impacts, illustrating the formats for use in the problems of Chapter 6, are used. In **Chapter 6,** an example of estimating the demand, physical and operational features, and the impacts associated with an HOV project are presented. *Following this, a number of variables including the diversion and modal shift are outlined to form the basis for alternative projects, each intended to be done by one or two participants. As with the long-range projects discussed above, the intermediate results and a range of final impacts for each project are provided in an* **instructor's manual.**

The level of detail of the projects is consistent with the "functional" design level used as a basis for decision making on project scope, impact estimation, and strategic planning, rather than on detailed design. This implies the determination of major dimensions, route layout, number of lanes, and location of stations, but not details such as lane widths, dimensions of curbs, signal supports, and position of signage—these latter are readily obtainable based upon federal, state, and industry standards.

Units of the international system (SI) and the English system are presented in the earlier chapters and where comparisons may be useful. However, the projects in Chapters 4 and 6 are conducted using the SI units only.

The book is short, because it is intended to illustrate the major features of the planning and design *process* of generating successful, pragmatic designs. It does not attempt to describe all of the methods, data, and nuances found in the extensive number of transportation project types that occur in practice. The reader should refer to the many excellent textbooks and technical literature for more detailed information. Quantitative input values and planning methods are being constantly updated, adopted, and documented by governmental and other organizations. Nevertheless, the process of planning, designing, and evaluating projects as a tool for decision making and strategic planning will undoubtedly exist for the foreseeable future. It is hoped that this book will assist readers in carrying out this work.

Comments and suggestions by students and colleagues, partially in response to earlier editions of this book and also to some related material that was used separately in course work and other presentations, have been much appreciated. Professor Jon Fricker of Purdue University, and William Hoey of California State University, made valuable suggestions and comments. Herbert S. Levinson contributed his many years of experience, livened by his imaginative approach to transportation and urban design problem solving, and made many aspects of this book more practical and readable. David Navick, doctoral student at Northeastern University, made a valuable contribution in checking the numerical examples. The style and presentation were aided considerably by Mary Ann Cottone, Margaret Cummins, and Steve Yun at Chapman and Hall.

Torrey Lee Adams made a number of helpful comments that contributed considerably to the clarity of the material. All errors and omissions are, of course, my own. I would be most grateful if readers would tell me about them and would welcome suggestions for possible improvements.

John G. Schoon
August, 1995

Transportation Systems and Service Policy

A Project-Based Introduction

PART 1

Introduction

1

Introduction to Transportation Planning and Policy

Major features of the traditional transportation systems planning process, official policy directed at attaining desirable goals and objectives through transportation actions, and the resulting impacts that help to define the success of the plans are briefly described in this chapter. Also, the distinction between long-range transportation planning and shorter-range transportation demand management actions and the characteristics of transportation corridors are outlined to establish a perspective on the scale and nature of the projects described in the ensuing chapters in Parts 2 and 3.

The Transportation Planning Process

To implement transportation plans that reflect societal values and the relevant physical constraints requires a set of goals reflecting that society's values which can be distilled to provide quantitative guidelines and parameters for planning and design. For the purposes of this book the guidelines that have evolved in general may be viewed in three ways:

- Goals and the planner's role
- Institutional involvement on transportation systems planning
- Current trends and problems affecting urban transportation.

Goals and the Planner's Role

In order to establish a working framework that reflects societal needs and articulates them in quantitative terms it is usual to establish goals reflecting that society's values. The goals may then be broken down into specific objectives, each quantified by appropriate performance criteria and numerical standards.

The framework may be illustrated in a simple way by considering the example of certain transportation-related actions to provide adequate access to health care:

- Societal *value:* Everyone's right to good health.
- One *goal* associated with maintaining good health: Ensure adequate transportation access to health care services.
- One *objective* appropriate to attaining the goal of ensuring access to health care facilities: Ensure that adequate bus service exists to provide transportation to health services.
- A performance *criterion* that reflects adequate bus access: Proximity of bus routes to health care services.
- A numerical *standard* appropriate for defining proximity of bus routes to health service facilities: Maximum route distance and bus stop location from each health service center is to be 0.25 km.

The need for mobility is essential. Yet other goals, such as those related to the maintenance of good health, result in the promulgation of numerical standards for air quality. For example, the maximum standard for primary carbon monoxide (CO) concentration during a 1-hour period is 40,000 mg/m^3 (Ambient Air Quality Standard, Environmental Protection Agency (1)).

Continually evolving societal perceptions and the increasing technical ability to measure the impacts of transportation actions require constant revision and reevaluation of a project's planning and design—particularly because current methods of prediction are approximate at best. In light of this, the planner and engineer increasingly require an appreciation of the process associated with conceptualization and eventual implementation of projects that may take over 20 years to bring to fruition. Manheim states that the role of the analyst (2) "is to intervene, delicately and deliberately, in the complex fabric of a society to use transport effectively, in accordance with other public and private actions, to achieve the goal of that society." Throughout the planning and design process, therefore, there is a continuing challenge to integrate the technical aspects of the work with emerging and sometimes poorly defined guidelines and with impacts that may not be fully understood. Clearly this demands of the planner and engineer a flexibility and a knowledge of the overall process, so that individual elements can be changed within a project, while leaving the whole intact and more responsive to emerging needs.

Institutional Involvement in Transportation Systems Planning—
A Response to Needs

Largely as a result of burgeoning population, economic expansion, and automobile ownership following the Second World War, formal, multimodal transportation planning in the United States was given considerable impetus by the Federal Aid Highway Act of 1962. This law required a continuing, comprehensive transportation planning process (known as the 3C process) to be carried out cooperatively by states and local communities. This process was particularly important because it recognized that all modes of transportation (not merely private vehicles on highways) should be included. It was followed at intervals by further laws related to transportation, environmental assessment requirements, and interagency cooperation. Based upon the 1975 Federal Highway Administration (FHWA) and Federal Transit Agency (FTA) (formerly the Urban Mass Transportation Administration [UMTA]) regulations, each Metropolitan Planning Organization (MPO) was required to include the following three elements in the preparation of a comprehensive transportation plan:

1. A long-range element with a time horizon of approximately 20 years
2. A transportation systems management (TSM) element—essentially a short-range element
3. A Transportation Improvement Plan (TIP) containing projects from 1 and 2 above for implementation in the next approximately five years, and that was responsive to documented priorities in the metropolitan area concerned.

The relationships of these elements to each other and further details of the process are shown in Figure 1-1.

The federal *Intermodal Surface Transportation Efficiency Act (ISTEA)* (3), enacted in 1991, retained the essential features of the 1975 regulations but provided greater flexibility. It provided additional funds for enhancing environmental-related measures and specified that the long-range planning requirements are to include land-use policies, intermodal connectivity, enhanced transit service, and management systems (1). The latter is a recognition that building new, extensive facilities to meet estimated demand may not be appropriate. The ISTEA authorized $151 billion over six years for highway, mass transit, and safety programs. The purpose of the act was set forth in its statement of policy: "It is the policy of the United States to develop a National Intermodal Transportation System that is economically efficient and environmentally sound, provides the foundation for the nation to compete in the global economy, and will move people and goods in an energy efficient manner."

Specific items in the ISTEA are excerpted below (1), along with (in italics) the ways in which the items relate particularly to transportation planning and design addressed in the ensuing chapters.

- A Congestion Mitigation and Air Quality Improvement Program was established. This sort of program must contribute to an area's meeting the

This chart shows the major activities in the transportation planning process. Each step is briefly discussed in this section.

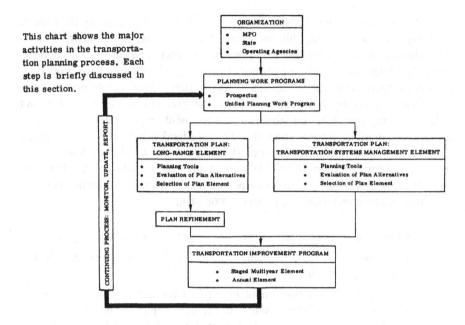

Figure 1-1 The long-range transportation planning process. 1975 **FHWA** and **UMTA** Regulations *Source*: Ref. (4)

Ambient Air Quality Standards (AAQS). *An essential output of the planning and design process described is the estimation of congestion indicators such as speed, vehicle volume, and travel time. Air pollution estimates are made for comparison with air quality standards.*

- Tolls were permitted on federal-aid highway facilities to a much greater degree than in the past. *Imposition of tolls on selected transportation facilities is related to service policy in that it affects the total user operating cost and, therefore, the selected mode of transportation, route, or time at which a trip is undertaken. These variables are included in the illustrative projects described in later chapters.*

- The metropolitan planning process was strengthened and the role of MPOs expanded in project selection and transportation decision making. MPOs continued to be required in all urbanized areas with a population of 50,000 or greater. *Transportation planning continues to be a process requiring consideration of individual projects, similar to those presented in later chapters, within the entire metropolitan area.*

- Urbanized areas with a population of more than 200,000 were designated as transportation management areas (TMAs). These areas had additional requirements for congestion management, project selection, and certification. *The use of high occupancy vehicle (HOV) facilities, discussed in Part 2, is one element of these requirements.*

• Each metropolitan area had to prepare a long-range plan, updated periodically, identifying transportation facilities that functioned as an integrated transportation system. It also included a financial plan to assess capital investment and other measures to preserve the existing transportation system and make the most efficient use of existing transportation facilities to relieve congestion, and indicated appropriate enhancement activities. A reasonable opportunity for public comment was required before the long-range plan was approved. In nonattainment areas, development of the long-range plan had to be coordinated with the development of transportation control measures for the state implementation plan (SIP) required under the Clean Air Act. *The focus on the planning and design process and projects in Parts 1 and 2 is on estimating the major impacts of alternative projects. This enables capital costs and effects on congestion, air pollution, and energy use to be estimated as a basis for public comment and compliance with air pollution requirements.*

• MPOs were required to include consideration of 15 interrelated factors in the development of their 20-year metropolitan transportation plan. One important factor was the effect of transportation decisions on land use and development and consistency with land-use and development plans. *The effects of land-use development in terms of population and employment changes in specific zones are shown in the illustrative project in Chapter 4.*

• In TMAs, the transportation planning process had to include a congestion management system (CMS) for the effective management of new and existing transportation facilities through the use of travel demand reduction and operational strategies. *The examination of travel speeds, volumes, and related performance measures described in the projects assists in quantifying the extent of congestion.*

• A Transportation Improvement Program (TIP) was required to be developed by the MPO in cooperation with the state and transit operators. The TIP had to be updated at least every two years and approved by the MPO and the governor, with a reasonable opportunity for public comment prior to approval. The TIP had to include a priority list of projects and a financial plan consistent with the funding that could reasonably be expected to be available. *An emphasis on the planning process and the need to establish analytical and design methods that are readily amenable to the updating process and investigation of alternatives are stressed throughout the book.*

As well as at the metropolitan level, states were also required to undertake a continuous *statewide* transportation planning process modeled on the metropolitan process. States were required to develop a long-range plan covering all modes of transportation, coordinated with the transportation planning carried out in metropolitan areas, with opportunity for public comment. The state plans and programs were to provide for the development of transportation facilities that functioned as an intermodal state transportation system. Integration of metropolitan and statewide plans may be included in federal major investment studies.

Trends in Urban Transportation—The Need for Planning

With the exception of temporary interruptions due to the energy crises of the 1970s, and despite improvements to transit systems, the general trend in urban travel has been a significant increase in automobile use and a decline in transit use—both in real and relative terms. This has led to increasing traffic congestion, with resulting adverse environmental and cost impacts. These combined effects have led to over 40 metropolitan areas in the United States exceeding the established Ambient Air Quality Standards. The most recent information on travel trends in urban areas is highlighted by the following data (5):

Place of residency and place of work relationships. The proportion of population living in metropolitan statistical areas (MSAs) grew from 77% to almost 80%. The proportion of those workers who worked in their MSA of residence remained unchanged at about 72%. Those working downtown declined in share, while those working in the remainder of the MSA and outside the MSA increased, in some cases significantly.

Transit use. Transit ridership remained at about 6 million riders from 1980 to 1990, declining only slightly in absolute terms by about 100,000 riders. As a result of the great increase in total commuters using automobiles the transit share declined from about 6.4% to about 5.3% of work travel. Detailed data shows the declines were all in bus travel, suggesting losses in the smaller metropolitan areas and shifts from bus to rail in the larger areas.

Modal trends. In terms of absolute amounts of travel, the drive-alone category again was the dominant gainer, *often increasing by more than the total increase in commuters* for the period because of shifts from other modes, as in Portland, Dallas, Houston, and the major parts of Seattle. In Buffalo, the total increase in driving also was more than double the increase in total commuters. Carpooling was mixed, predominantly showing actual declines but increasing in some areas, notably Los Angeles.

Vehicle use. Vehicles per household continued to increase despite steady decreases in household size. More daily trips and longer trip lengths resulted in a 29% increase (3.6% compounded annual increase) in daily household vehicle miles traveled from 1983 to 1990.

Trip purpose and length. Work trips continued to account for the largest proportion of household travel, both in terms of miles and in number of trips. Average vehicle trip lengths, which had been decreasing from 1969 to 1983, showed increases in 1990. The largest increase in trip length was in work trips.

Vehicle occupancy. From 1974 to 1990 the average vehicle occupancy, calculated as person miles per vehicle mile, declined steadily for commuting and shopping, from approximately 1.3 to 1.1. Several factors contributed to the general decline in vehicle occupancy, including the increased number of vehicles per household and the decrease in average household size.

Automobile ownership. The percentage of households without a vehicle dropped from 20.6% in 1969 to 9.2% in 1990, whereas the percentage of households with three or more vehicles available quadrupled. Over the 1969 to 1990 period, the total number of households increased by 49% while the number of household vehicles increased by 128%.

The continuing increase in use of private automobiles, due largely to increasing numbers of households, higher incidence of car ownership, and increased urban population dispersion, has resulted in adverse environmental impacts and underscores the need to plan on a multimodal basis. There is no guarantee that planning will result in reduced automobile travel. Nevertheless, the provision of facilities that contribute to the mobility offered by other modes is an essential feature of any transportation system that offers a balanced set of options for the user.

Distinction Between Long- and Short-Range Planning

Some overlap may exist between long-range plans and TSM actions, and some useful distinctions are as follows (6):

1. Whereas long-range planning deals with capital-intensive improvements requiring several years to plan and implement, TSM embraces low-budget actions of short durations (possibly a year or two to plan and implement).

2. Recycling of TSM projects during the long-range timeframe is quite possible.

3. TSM strategies may work out to be alternatives to long-range components.

4. "Major" TSM projects may be synonymous to "minor" capital-intensive projects, particularly at the middle range levels.

5. There is an inherent flexibility in TSM projects. Proper monitoring and effective feedback may result in appropriate modifications to the original project and may lead to overall economy.

Table 1-1   Key differences between TSM and long-range planning

	TSM	Long-range
Problems	Clearly defined, observable	Dependent on growth scenarios and projected travel
Scale	Usually local, subarea, or corridor	Usually corridor or regional
Objectives	Problem related	Broad, policy related
Options	Few specific actions	Several modal, network, and alignment alternatives
Analysis procedures	Usually analogy or simple operational relationships	Based on trip and network models
Response time	Quick response essential	Not critical
Product	Design for implementation	Preferred alternative for further study or detailed design

Source: Ref. (6)

A further clarification of the differences between the two types of planning is provided in Table 1-1. Several of the major differences will be addressed in the background and formulation of the short-range projects described in later chapters, and greater detail of TSM is provided in (7) and (8).

It should be noted that for the ISTEA the term transportation demand management (TDM) has replaced TSM, although the essential features, for the purposes of this book, are the same. For consistency, we use the term TDM from this point on, except where TSM is used in a specific reference or document.

The Planning Framework

Within the framework of continuing, comprehensive, long-range urban transportation systems planning and development, project planning generally involves the planning, functional design, and evaluation of highway, public transportation, parking, terminal, and related projects, often of an intermodal nature.

Examples of typical projects of the program include multimodal transportation corridors (the subject of the projects in this book), transportation systems management (TSM) improvements and/or transportation demand management (TDM) projects, street and traffic control projects, public transportation service improvements, and parking, pedestrian, and terminal facilities. Examples of various categories of urban transportation projects are shown in Figure 1-2.

In practice, the long-range transportation plan is usually tied to a five-year state or metropolitan transportation improvement program. The separation in this book of long- and short-range planning and design recognizes this. However, the book

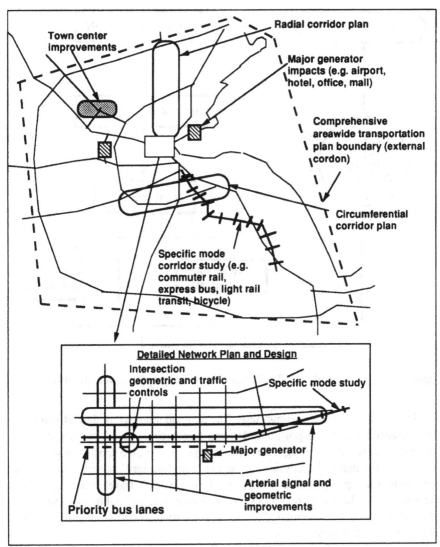

Figure 1-2   Examples of transportation project locations and extent

is divided into two parts to reflect the fact that current analytical procedures have evolved somewhat separately for the two planning horizons. Also, the focus on long-range planning in Part 2 is intended to provide the perspective on the total system before Part 3 is undertaken.

An overview of the demand estimation process described in later sections is presented in Figure 1-3, indicating the major areas of demand and supply, and the

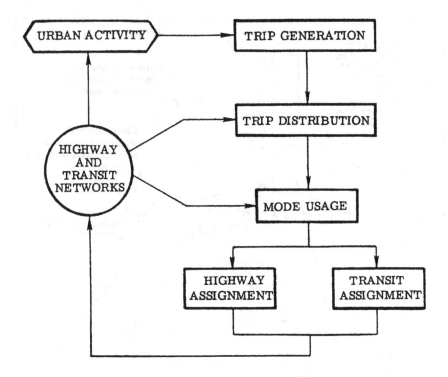

Figure 1-3   Demand analysis—The forecasting model system   *Source*: Ref. (4)

relationship between the various steps. The demand analysis method is similar to the sequential urban transportation demand forecasting procedure typically used by federal, state, and other agencies for strategic, long-range urban transportation planning. This demand analysis is done using preliminary characteristics of the transportation supply, such as estimates of approximate travel times, and using the assumption that the basic modes will be automobile and public transportation. This, however, does not preclude investigation of pedestrian, bicycle, or other modes where warranted.

Following determination of the functional design in greater detail, including inputs such as modal split, level of service, and transit performance characteristics, the impacts are estimated based upon unit values of capital costs, energy use, and air pollution. When the evaluation of the impacts has been completed and the results examined, it is often necessary to recycle and modify the supply and demand inputs. When the results of the evaluation are satisfactory, or the candidate options have been screened to a reduced number, the desired alternatives can be reviewed and selected and then refined to the desired level of detail. This refinement is not part of the projects described in detail here, yet could be conducted if required, using the same or a modified procedure.

Level of Detail

The process illustrated earlier in Figure 1-3 is essentially similar to the demand estimation process of Figure 1-3. The level of detail of the project is mostly at the "sketch" level shown in Figure 1-4. The different levels of planning shown in this figure differ in complexity, cost, level of effort, sophistication, and accuracy, but each has its place in travel forecasting. The levels are explained (4) as follows:

Sketch Tools. Sketch planning is the preliminary screening of possible configurations of concepts. It is used to compare a large number of proposed policies in enough analytical detail to support broad policy decisions. Useful in both long- and short-range regional planning and in preliminary corridor analysis, sketch planning, at minimum data costs, yields aggregate estimates of capital and operating costs, patronage, corridor traffic flows, service levels, energy consumption, and air pollution. The planner usually remains in the sketch planning mode until completion of comparisons of possibilities or until a strategic plan worthy of consideration at a finer level of detail is found.

The sketch planning mode is useful also in that it enables "fatal flaws" in a proposed alternative to be identified. For example, if a proposed plan indicated an air pollution level greatly in excess of that allowable, based on the sketch planning level of detail, it could be discarded from more detailed investigation or considerably modified before time and resources were spent on its further examination.

Traditional tools. Traditional tools treat the kind of detail appropriate to tactical planning; they deal with many fewer alternatives than sketch tools, but in much

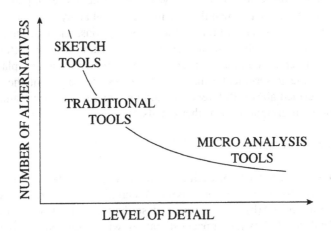

Figure 1-4 Levels of planning detail   *Source*: Ref. (4)

greater detail. Inputs include the location of principal highway facilities and delineated transit routes.

At this level of analysis the outputs are detailed estimates of transit fleet size and operating requirements for specific service areas, refined cost and patronage forecasts, and level-of-service measures for specific geographical areas. Household displacements, noise, and aesthetic factors can also be evaluated.

The cost of examining an alternative at the traditional level is 10–20 times its cost in sketch planning, although default models, which dispense with many data requirements, can be used for a less expensive first look. Promising plans can be analyzed in detail, and problems uncovered at this stage will suggest a return to sketch planning to accommodate new constraints.

Microanalysis tools. Microanalysis tools are applicable as the time to implement a project grows near. They are the most detailed of all planning tools. At this level of analysis, one may wish, for example, to make a detailed evaluation of the extension, rescheduling, or repricing of existing bus service; to analyze passenger and vehicle flows through a transportation terminal or activity center; or to compare possible routing and shuttling strategies for a demand-activated system. Final analysis at this level is prohibitively expensive except for subsystems whose implementation is very likely, and whose design refinements would bring substantial increases in service or significant reductions of cost. It is most effective in near-term planning when a great many outside variables can be accurately observed or estimated. It is sometimes necessary, however, to use microanalysis tools to supplement the output of traditional longer-range planning.

In the examples described later, the input data (including the policy variables) are provided for the design, or horizon, year at the sketch level of planning detail. Furthermore, it is also assumed that the mathematical analysis models have been calibrated where necessary to reflect these assumptions, and that the transportation supply and unit values of impacts are at the same level of detail. A wide range of commercial computer programs are currently in use by planners and engineers in the governmental and private sectors to carry out the numerical analyses described above. Reference should be made to specific manufacturers' and suppliers' information for further details.

Impacts and Evaluation

Estimates of the impacts of each alternative are necessary for conducting the evaluations as a basis for decision making. A summary of this process is shown in Figures 1-5(a) and 1-5(b). In the projects described later, the two elements in this process, facilities design and impact estimation, will be the focus of our efforts.

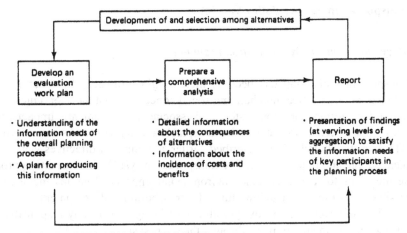

(a) SCOPE OF THE EVALUATION PROCESS

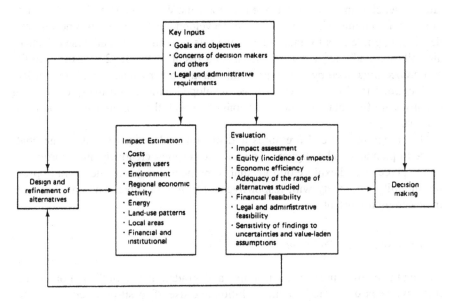

(b) CONCEPTUAL FRAMEWORK FOR EVALUATION

Figure 1-5   Aspects of project evaluation   *Source*: Ref. (7)

Transportation Service Policy Variables

Policy Directions in Transportation Planning

Transportation planning has become increasingly influenced by federal, state, and local policies reflecting changing social concerns about the economic effects of an efficient transportation system and associated environmental impacts, both of which are seen as essential to an acceptable quality of life. As a result of the ISTEA in the United States described earlier, a number of directives were made to encourage and require that future transportation systems incorporate appropriate measures to reduce adverse environmental impacts. Penalties included the possible loss of federal matching funds for transportation development, as well as other loss of revenues, and the possibility of being held legally responsible for adverse environmental impacts through the court system.

Globally, increasing environmental degradation has resulted in several multilateral agencies, including the United Nations, the World Bank, and the Organisation for Economic Co-operation and Development (OECD), promoting and developing guidelines for taking an active role in reducing the adverse effects of air pollution and energy consumption. As a result of the 1992 "Earth Summit" conference organized by the United Nations and other agencies and held at Rio de Janeiro, Brazil (10), several of the developed nations and many developing countries are formulating guidelines consistent with the conference's recommendations.

The foregoing societal pressures and institutional actions intensify the challenge to planners and designers. The need to respond creatively and thoroughly, and to perceive and explore options that combine technical excellence in improving mobility and safety with responsiveness to less tangible concerns, becomes increasingly important.

Energy Use and Air Pollution

Among the many sources of environmental degradation, two of the major sources that have been quantified are air pollution and use of fossil-powered energy for transportation. Essentially, it is the conversion of fossil energy by means of our current technology in order to experience desired economic and social benefits that generates much of the source of environmental degradation. To the extent that communities can attain the same or a better standard of living, while at the same time making more efficient use of energy and reducing air pollution, so the possibilities for improving environmental quality (or reducing the rate of its decline) will exist.

In order to attain improved mobility while reducing adverse impacts by planning and designing the physical and operational features of the transportation system, the methods and example problems described later will focus on two primary concerns: energy use and air pollution. Within these two areas, the levels of the input variables that may be established by government agencies will provide a basic means of exploring the likely impacts.

Although many other requirements and concerns are of key importance in formulating desirable transportation projects (some of which are listed in Table 1-2), these two, and the requirement to manage demand by employing greater effi-

Table 1-2   Metropolitan Transportation Planning Factors

1   Preservation of existing transportation facilities and, where practical, ways to meet transportation needs by using existing transportation facilities more efficiently.
2   The consistency of transportation planning with applicable federal, state, and local energy conservation programs, goals, and objectives.
3   The need to relieve congestion and prevent congestion from occurring where it does not yet exist.
4   The likely effect of transportation policy decisions on land use and development and the consistency of transportation plans and programs with the provisions of all applicable short- and long-term land use and development plans.
5   The programming of expenditures on transportation enhancement activities as required in section 133.
6   The effects of all transportation projects to be undertaken in the metropolitan area, without regard to whether such projects are publicly funded.
7   International border crossings and access to ports, airports, intermodal transportation facilities, major freight distribution routes, national parks, recreation areas, monuments, historic sites, and military installations.
8   The need for connectivity of roads within the metropolitan area with roads outside the metropolitan area.
9   The transportation needs identified through use of the management systems required by section 303 of this title.
10   Preservation of rights-of-way for construction of future transportation projects, including identification of unused rights-of-way which may be needed for future transportation corridors and identification of those corridors for which action is most needed to prevent destruction or loss.
11   Methods to enhance the efficient movement of freight.
12   The use of life-cycle costs in the design and engineering of bridges, tunnels, or pavement.
13   The overall social, economic, energy and environmental effects of transportation decisions.
14   Methods to expand and enhance transit services and to increase the use of such services.
15   Capital investments that would result in increased security in transit systems.

Source: Ref. (3), reported in Ref. (11)

Table 1-3 Energy Inputs for Various Urban Transportation Modes

Transport mode	Energy input	
	BTU/Passenger ml	kJ/Passenger km
Bicycle	150	98
Walking	400	262
Rail transit	1,650	1,082
Commuter railroad	2,500	1,639
Bus transit	3,190	2,092
Automobile (general average)	6,840	4,485
Small car (all trip purposes)	2,620	1,718
Large car (all trip purposes)	5,100	3,344

SI to English conversion factors: 1 Btu/Passenger ml = 0.6557 kJ/Passenger km

Source: Based upon data in Ref. (12)

Note: Actual energy input values will depend extensively on vehicle occupancy and trip characteristics

ciency in the use of facilities, are most clearly and directly related to the facility's design. The sources of current guidelines regarding energy consumption and air pollution concerns in the formal planning process in the United States are described below (11).

Energy Consumption

Following earlier environmental regulations, the National Energy Conservation Policy Act of 1978 required states to undertake specific conservation actions including the promotion of car- and vanpools and other conservation efforts. Resulting regulations in 1980 required that all phases of transportation projects from planning to construction and operations be conducted in a manner that conserves fuel. They incorporated energy conservation as a goal of the urban transportation planning process and required an analysis of alternative transportation systems management (TSM) improvements to reduce energy consumption. By examination of Table 1-3, it can be seen that considerable savings in energy use are likely to accrue if vehicle miles traveled (VMT) by automobiles can be reduced. Energy conservation has become integrated with the urban transportation planning process as a result of federal and state legislation and regulation.

Air Pollution

After passage of the Clean Air Act Amendments of 1970, average automobile emissions dropped from 85 grams per mile of carbon monoxide (CO) in 1970 to

25 grams per mile in 1988. Lead usage in gasoline dropped by 99% between 1975 and 1988. From 1978 to 1988, transportation-related emissions decreased 38% for CO, 36% for hydrocarbons, and 15% for nitrogen oxides (NOX). All occurred despite a corresponding 24% increase in VMT during the same period. Nevertheless, by 1988, nearly 150 urban areas failed to meet the "nonattainment areas" (NAAQS). These areas had to undertake varying degrees of measures to reduce air pollution levels for specified emissions to an acceptable level. These measures included implementation of projects that would tend to reduce emissions, such as transit improvements, high occupancy vehicle (HOV) lanes and other HOV incentives, traffic flow improvement, fringe parking, single occupant vehicle disincentives, including pricing, and incident management. Many of these measures assist in reducing traffic congestion and resulting low traffic speeds. The effects on most air pollution emissions of increasing speed can be seen in the graphs of Figure 1-6.

Policy Variables Affecting Energy Use and Emissions Levels

Role of Public Policy in Shaping the Transportation System

Within certain limits, the variables affecting demand for and supply of transportation can be controlled by the governmental agency having jurisdiction over the transportation system. Major elements of transportation supply have been identified that may be influenced by various policies in order to attain certain objectives (12). These include service frequency and cost to users, characteristics of infrastructure and operation, including service subsidies. Although the effectiveness in achieving the objectives is often limited because of the effects of market and political forces that are difficult to predict, exploration of the effects of different levels of variables (sometimes referred to as scenarios) often provides valuable insights into costs and impacts such as air pollution and energy use. This is an essential feature of the planning process.

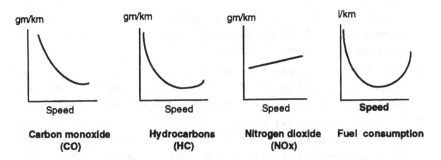

Figure 1-6 Emissions and fuel consumption for highway vehicles   *Source*: Based upon data in Ref. (12)

Selection of Policy Variables

As indicated above, demand for transportation can be affected by a wide range of policy techniques. These techniques may be further broken down into subcategories: transportation demand, transportation supply, and nontransportation variables. *The former two are usually related to the service level or generalized cost of using the system as perceived by the individual at the time of use.* They are the variables selected for use in the examples in later chapters to illustrate their effects on the various plan and design proposals.

The alternative transportation plans and designs that may result from the imposition of service policy variables typically range from "auto intensive" to "transit intensive" in comparing how the capital costs and environmental impacts vary for each alternative. In practice, a final decision regarding the actual plan to be implemented often depends upon considerable investigation of, and compromise between, policies, following a detailed evaluation. Also, the policy variables are often applied incrementally because of the time taken to coordinate the implementation process from concept to legislative action. The various policy alternatives, as well as a brief note about how they may be implemented, may therefore comprise different values of variables. Some of these include:

1. Service time for transit users (walking, waiting, transfer time), which may be varied by increasing the transit vehicle frequency, the route, or timing. Timing in particular can affect the total time spent at transfer points between routes. The lower the service time, the greater the tendency for the use of public transportation.

2. Cost differentials in private vehicle use versus transit use, which may be varied by techniques such as parking surcharges, tolls, and transit fare subsidies. Congestion pricing to discourage travel by private automobile during peak periods is also being considered in a number of locations. Usually, cost differentials are expressed in terms of "out-of-pocket" costs to each user.

3. In-vehicle time differentials, which may be reduced by provision of high occupancy vehicle (HOV) lanes and ramps and other access priority methods, thus encouraging their use.

4. Vehicle occupancy, which may be varied by the extent of incentives for carpooling, vanpooling, and other shared ride techniques. Increasing vehicle occupancy encourages the movement of people rather than vehicles.

5. Level of service (LOS). Specification of the design LOS "E" (capacity conditions) instead of "C" or "D" for limited access highways in peak hours has the effect of increasing the travel time for private motor vehicle users during peak hours, thereby making the use of alternative modes relatively more attractive.

Other methods of discouraging private automobile use, although maintaining or improving overall mobility, would include increasing parking costs at downtown locations while simultaneously increasing the supply of affordable parking at outlying parking lots along transit lines. Of the nontransportation policies, implementation of staggered work hours, flex time, and four-day work-weeks are examples. Numerous other methods of encouraging transit use and discouraging single occupancy automobile work trips during peak hours exist. Many of these are described in (8), and include implementation of:

- Fixed route local and express bus service to improve accessibility by transit vehicles
- Paratransit service such as minibuses where fixed route transit vehicles are uneconomical
- Land-use policies for improved transit access
- Site design criteria to increase transit access
- Transit-oriented parking management strategies to provide parking spaces along transit lines
- Employer initiatives to encourage transit use
- Strategic approaches to avoiding congestion, including growth management, road pricing such as tolls, auto restricted zones, parking management, site design to minimize traffic, demand management agreements, ridesharing, alternative work hours, and trip reduction ordinances.

From the transit users' point of view, each of the above actions translates into faster travel times and/or reduced travel cost. These are stated in quantitative terms in the policy variables described later in this chapter.

As well as the established policy variables, other inputs to the planning process may differ from the predicted values because of inability to control or model them with sufficient accuracy. Two of the most difficult inputs to control and, hence, to predict, are the population level of each zone, and the spatial distribution of new residential and industrial/commercial development. Although actual control over the levels of these variables has often proved to be extremely difficult in North America, the evolution in computing methods using mini- and microcomputers has assisted in incorporating these variables quickly and with minimal effort into the planning process, thereby improving the ease and speed of analysis in response to changing conditions.

Transportation Corridor Planning and Design

Transportation corridor analysis and planning is a commonly occurring element of total transportation planning in most cities. In a wider context, a comprehensive, areawide transportation study may consist of the analysis of several corridors. At a subcorridor level, consideration of the private and public transportation modes in portions of corridors often occurs during the analysis of new traffic generators and upgrading of transportation facilities.

Planning and design of transportation facilities in major corridors, therefore, provides the opportunity for analyzing and examining the major features of most metropolitan and urban area transportation requirements. The exact boundaries of each corridor are often difficult to define precisely, yet the location of the major transportation facilities themselves and the land uses that they serve enable a reasonably well-defined boundary to be identified.

Characteristics of Transportation Corridors

Although transportation corridors are evident in almost all urban and regional areas, a city of population greater than approximately 50,000 will usually have two or more corridors depending on its location and the extent of highways, railroads, and, perhaps, waterborne transportation serving it. Most multimodal corridors may be typified by the following characteristics:

1. The spine usually comprises major highway and/or public transportation routes extending radially between the outer suburbs and the city's center, although circumferential corridors may also be defined.

2. Most multimodal transportation corridors have evolved along the path of least physical resistance and are, therefore, often located along or parallel to rivers or the shores of other bodies of water.

3. Examination of radial highways and the location of fringe and suburban communities will enable an approximate dividing line between one corridor and the next to be identified.

4. Communities that are a part of each radial corridor will use the transportation facilities in each one for the majority of trips to and from the central city area.

5. Feeder services for public transportation may be available, and local bus and other transit services will connect into the major radial transportation facilities.

6. Although congestion and inadequacy of facilities may be evident at various points along the corridor, the concentration of deficient capacity will usually be most evident at the portion of the corridor immediately outside of the central city area. This is because feeder and other routes and the radial highways and transit facilities are constrained in this area by restricted availability of land, extensive local traffic, often including freight and other commercial vehicles, and congestion in the central area radial routes.

7.   Parking at the city's fringe area and suburban transit lots must be closely coordinated with parking policies for central areas and placement of parking facilities so that excessive traffic from the corridor is not diverted to central area streets. Land uses along the corridor will generate movements between outlying locations and the downtown. Therefore, coordination of land uses and the associated transportation systems is essential in the planning effort.

Most cities may be divided into corridors of travel, particularly cities that exhibit a strong center with radial configuration of highways and public transportation routes. An example of the corridor structure for a portion of the Boston, Massachusetts, metropolitan area is shown in Figure 1-7. The figure also indicates the volume of passengers approaching the city's center in each of the six major travel corridors. The volume of persons entering the city area (including those who may also pass through to other destinations) varies between about 20,000 and 60,000 during the period 7:00 A.M. to 6:00 P.M. This corresponds with peak hourly volumes of between about 10,000 to 30,000 in one direction during each A.M. and P.M. peak hour.

Greater detail of the modes and morning peak hour passenger volumes in Corridor 6 in the Boston area is shown in Figure 1-8. In this particular corridor, over 30,000 passengers are estimated to travel into the downtown area during the A.M. peak hour, using local buses and cars on surface arterial streets; cars, vans, and commuter buses on a freeway; a rail rapid transit line; commuter rail line; and passenger ferries. Other corridors feature different mixes of these and other modes, including a small proportion of commuters using bicycles. Table 1-4 summarizes the principal features of selected public transportation modes employed on major urban corridors.

Review

This chapter has outlined very briefly the elements of transportation systems planning that are most relevant to the quantitative planning and design tasks described in the ensuing chapters. The ways in which this and the following chapters relate is summarized below.

The transportation planning process. The overview of current legislation concerning the objectives and process of transportation planning and the way long- and short-range plans form a part of the total plan establishes a framework within which most planning and design efforts are conducted and for which the results must be consistent. The distinction between the levels of detail of the various planning tools, ranging from the sketch level to detailed operational planning, indicates the type of modeling typically undertaken in practice. The

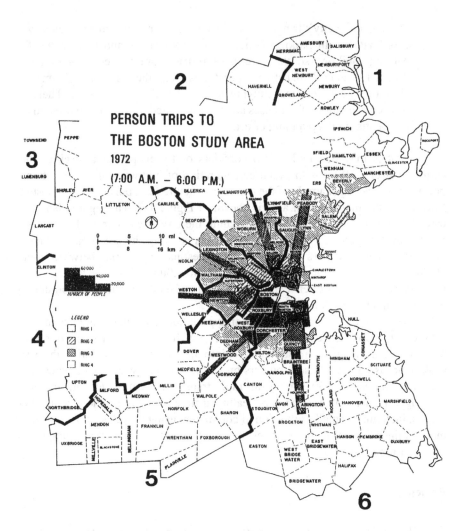

Figure 1-7   Transportation corridors in the Boston, Massachusetts, area   *Source*: Ref. (14)

emphasis throughout the remaining chapters will be on the sketch planning level of detail.

Impacts and evaluation. For a proposed plan and preliminary design to satisfy the requirements for increasing mobility with acceptable levels of impacts, it is essential that the kinds of impacts be understood and be readily quantifiable. The brief introduction and background presented illustrate some of the major

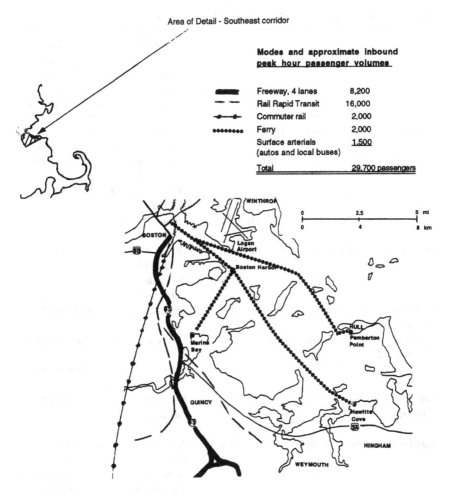

Area of Detail - Southeast corridor

Modes and approximate inbound peak hour passenger volumes

▬▬▬	Freeway, 4 lanes	8,200
— —	Rail Rapid Transit	16,000
—●—●—	Commuter rail	2,000
●●●●●●●	Ferry	2,000
	Surface arterials (autos and local buses)	1,500
	Total	29,700 passengers

Figure 1-8 Details of transportation modes in the Southeast Corridor, Boston, Massachusetts

considerations and trends in energy use and air pollution related to transportation. More detailed data and its application to specific examples will be addressed later.

Policy variables. The imposition of specified levels of the variables that can affect the amount of travel and the mode used is one of the ways in which public agencies can guide compliance with regulations affecting provision of mobility and the extent of impacts. Although many different policy variables may be identified, the ones selected in this case for application are service time for transit users, cost differentials in private vehicle use versus transit, in-vehicle time differentials, vehicle occupancy, and level of service. Throughout

Table 1-4 Characteristics of Selected Urban Transportation Modes

Mode	Typical way dimensions	Vehicle capacity	Vehicles per train	Typical operating speed (urban) km/h	Typical hourly passenger volumes (passengers one-way per hour)
Bicycle	1.5 m (1-way)	1	1	10 to 24	300 (Varies widely)
Automobile	3.7-m lane	4 to 6	1	On freeways 48 to 100 On surface streets 15 to 48	2,000 to 4,000
Bus, local	3.7-m lane	40 to 60	1	On surface streets 15 to 48	2,000 to 3,000
Bus, express	3.7-m lane	40 to 50	1	On freeways 48 to 100 On surface streets 15 to 48	3,000 to 5,000
Bus Rapid Transit	3.7-m lane	40 to 50	1	On separate lane 48 to 100	4,000 to 8,000
Light Rail Vehicle (LRV)	10 m (2-way)	150	2 to 4	On separate way 25 to 60 On city streets 15 to 30	5,000 to 15,000 3,000 to 8,000
Rail Rapid Transit (RRT)	10 m (2-way)	150	4 to 10	48 to 100 (always separate way)	10,000 to 40,000
Regional Rail	40 m (2-way)	150	4 to 10	48 to 100 Way may be shared with intercity rail	10,000 Depends extensively on scheduling

the remaining sections, specific aspects of the demand and supply analyses will be described on the assumption that the methodology must be responsive to these variables.

Transportation corridor planning and design. The transportation corridors of major cities exhibit a number of common characteristics and are often selected as the study area for long- and short-range transportation improvements. Because of this, the description of the major characteristics that typify many corridors will provide a basis for the example studies and for the projects for solution in Chapters 4 and 6.

References

1. Environmental Protection Agency, *Federal Register,* 36, No. 84, Washington, D.C., April 30, 1971.

2. Manheim, M., *Fundamentals of Transportation Systems Analysis,* MIT Press, Cambridge, MA, 1985.

3. U.S. Department of Transportation, *Intermodal Surface Transportation Efficiency Act,* Washington, D.C., 1991.

4. U.S. Department of Transportation, Federal Highway Administration, Urban Mass Transportation Administration, *Urban Travel Demand Forecasting—A Self Instructional Text,* Washington, D.C., 1977.

5. U.S. Department of Transportation, *Transportation Statistics,* Washington, D.C., 1994.

6. Transportation Research Board, *Transportation System Management,* Special Report 172, National Research 153, TRB, Washington, D.C., 1977.

7. Transportation Research Board, *Simplified Procedures for Evaluating Low-Cost TSM Projects—User's Manual,* NCHRP Report 263, TRB, Washington, D.C., October 1983.

8. Institute of Transportation Engineers, *A Toolbox for Alleviating Traffic Congestion,* Washington, D.C., 1989.

9. U.S. Department of Transportation Federal Highway Administration, *Transportation Planning and Problem Solving for Rural Areas and Small Towns, Student Manual,* Washington, D.C., 1983.

10. United Nations Centre for Environment and Development, *Agenda 21,* New York, 1992.

11. Weiner, E., *Urban Transportation in the United States,* U.S. Department of Transportation Technology Sharing Program, Washington, D.C., 1992.

12. Homburger, Wolfgang S., and James H. Kell, *Fundamentals of Traffic Engineering,* 12th ed., University of California at Berkeley, 1988.

13. Meyer, Michael D., and Eric J. Miller, *Urban Transportation Planning. A Decision-Oriented Approach,* McGraw-Hill, New York, 1984.

14. Wilbur Smith and Associates, "An Access Oriented Parking Strategy for the Boston Metropolitan Area," prepared for Massachusetts Department of Public Works and the U.S. Department of Transportaion, Federal Highway Administration, Boston, 1974.

Other relevant texts and documents are listed in the Bibliography.

PART 2

Long-Range Transportation Planning

2

Transportation Demand Analysis

This chapter provides an overview and examples of transportation demand analysis, from specification of the characteristics of the urban area in which the project is located to estimation of the vehicle volumes on individual links of the network. It describes a basis for the data selected and the methods used in Chapter 4, where a plan, design, and impact estimation procedure is conducted in the example project.

The Planning Process

The major features of the planning process were described in Chapter 1. The main points of the urban activity and transportation demand analysis phases described in this chapter are summarized as follows:

Urban Activity

- Urban area demographics, primarily population and income.
- Topography, including natural landforms and man-made features, and areas of environmental value, such as wetlands, and historic areas.
- Economic activity, including industry, business, retail, and transportation facilities.

Transportation Demand Analysis

- Trip generation—the number of trips by persons and vehicles.
- Trip distribution—the distribution of the trips throughout the area.
- Modal split—the type of mode, usually private or public, used for the trip.
- Traffic assignment—the highway or transit route that the trip takes.

The analysis methods used for estimating the outputs resulting from the demand analysis, which comprises a part of the overall planning process, have been selected from several sources used by transportation planning agencies. See, for example, (1). The methods selected are presented in a form that allows the estimations to be performed manually or by means of computer worksheets.

Urban Activity

Background

Because transportation is regarded as a "derived" demand, it follows that the geographic and demographic characteristics of the area will be instrumental in shaping major features of the transportation demand and supply. The urban activity is usually shaped by regional and national economic forces and trends, as well.

Land Use and Transportation Interaction

The land-use and urban activity process has been summarized by Manheim (2) in Figure 2-1, illustrating that the traffic flows are shaped by the activity system, which in turn is affected by the new traffic flows, resulting in a dynamic interaction between the activity and the transportation system. In these notes, we consider the activity system and the resulting person and vehicle flows fixed as of a "design" or "horizon" year, knowing that any plans must be continuously updated and revised to suit changing activities and features of the transportation system.

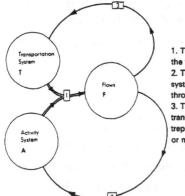

1. The flow pattern in the transportation system is determined by both the transportation system and the activity system.

2. The current flow pattern will cause changes over time in the activity system: through the pattern of transportation services provided and through the resources consumed in providing that service.

3. The current flow pattern will also cause changes over time in the transportation system: in response to actual or anticipated flows, entrepreneurs and governments will develop new transportation services or modify existing services.

Figure 2-1   Relations between transportation and activity systems   *Source*: Ref. (2)

Data Needs

Pertinent characteristics of the urban area and the transportation system must be specified or otherwise determined because they will affect the demand for transportation and the opportunities and constraints on designing the system. Urban activity and topographic conditions will affect the demand, supply features, and impacts of a proposed plan.

Demand for transportation will be affected by the extent of the urban area, the spatial layout, density, and the locations of settlements, and their residential, work, recreational, and institutional components. As well as the central business district (CBD), travel generators such as industrial complexes, airports, and sports facilities must be included. Also, the incomes of the residents and funds available within the region for economic and infrastructure development will play a key role in establishing the level of demand. In general, mobility increases with income.

The capacity, speed, and vehicle characteristics of the transportation modes that are compatible with the demands, and the transportation supply (including the available rights-of-way) will be affected by the area's topographic features. These include the existing urban area and regional transportation routes, development density, environmentally sensitive areas, and constraints on route location such as habitation, steep terrain, and bodies of water (although the last item may sometimes offer opportunities for travel as well as being a constraint).

Regarding transportation impacts, the climatic, topographic, and cultural characteristics of the area as well as regulatory and other requirements that may be unique to a specific area will provide primary inputs to the estimation of impacts.

Topographic maps and tabulations showing summaries of demographic information, such as population, income, household size, automobile ownership, and employment, as well as information on activity centers that generate extensive transportation demand provide useful information.

An example of a portion of a study area divided into traffic analysis zones and showing some of the data mentioned above is shown in Figure 2-2. The zones usually represent areas of relatively homogeneous characteristics such as household income, but the boundaries may also reflect the presence of significant topographic features such as rivers or freeways. An inventory of existing transportation system information is also necessary, including route location, and physical and operating features, such as speed, level of service, operating and maintenance costs, and current capital expenditures. Zone plans differ for different types of studies, with sketch plans having fewer and larger zones, or districts.

Other sources of data on most urban areas are available. In addition to census data, states, cities, and transportation planning agencies maintain records of the needed information. The data are also available in digitized format from the U.S. Bureau of the Census. An example of the type of data available for various census tracts in Massachusetts is shown in Figure 2-3.

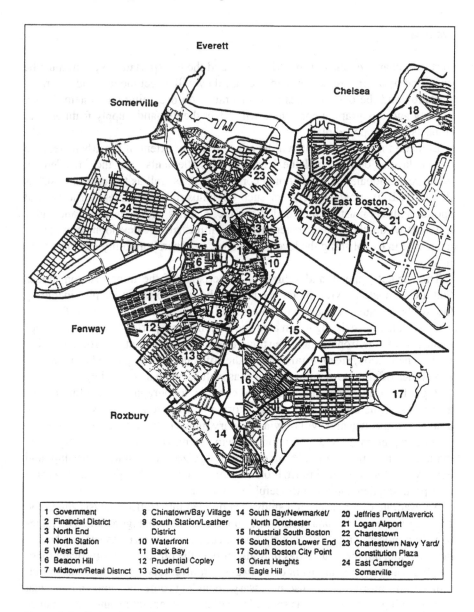

1 Government	8 Chinatown/Bay Village	14 South Bay/Newmarket/	20 Jeffries Point/Maverick
2 Financial District	9 South Station/Leather	North Dorchester	21 Logan Airport
3 North End	District	15 Industrial South Boston	22 Charlestown
4 North Station	10 Waterfront	16 South Boston Lower End	23 Charlestown Navy Yard/
5 West End	11 Back Bay	17 South Boston City Point	Constitution Plaza
6 Beacon Hill	12 Prudential Copley	18 Orient Heights	24 East Cambridge/
7 Midtown/Retail District	13 South End	19 Eagle Hill	Somerville

Figure 2-2   Example of traffic analysis zones   *Source*: Ref. (3)

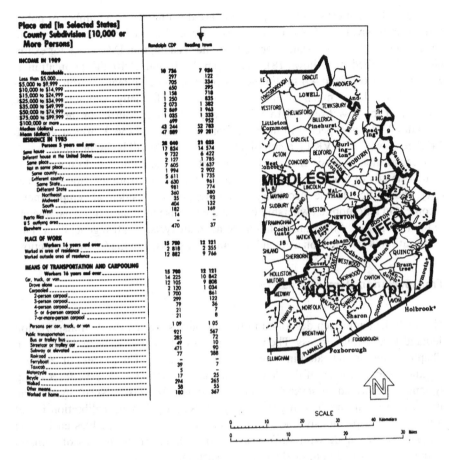

Figure 2-3 Example of transportation and demographic data by census areas
Source: Based upon Ref. (4)

Analysis Methods and Links to Other Phases

Many methods of estimating future land use and urban activity exist. They range from extensions of zoning regulations that place restrictions on the type and density of land uses at various locations and, therefore, imply a total holding capacity, to mathematical models that include consideration of regional and urban economic development and environmental protection and enhancement. Descriptions of these models are beyond the scope of this book.

Physical and demographic patterns of urban areas related to transportation are complex and vary with terrain, economic development, technological capability, and cultural determinants. For urban areas in North America, patterns of development related to transportation have been described in detail. The relationships between urban patterns and transportation are documented in a number of publications, extensively in (5, 6) and, more recently, in (7). The land-use development pattern related to transportation demand in the Washington, D.C., area, for example, is described as follows (8): "Between 1990 and 2010 the population is expected to grow from 4.4 million to 5.4 million and employment from 2.7 million to 3.8 million (much of this in the suburbs). The daily commute trips are anticipated to grow from 4.0 to 5.7 million, with three-fourths in the suburbs. Weekday VMT is expected to increase from 9.6 million (156.0 million VKT) to 147.1 million (237.0 VKT); again much of this growth will be in the suburbs."

Because the emphasis here is on transportation planning, we do not include further details of land-use planning models. The outputs from the land-use planning phase include the necessary quantification of residential, employment, income and automobile ownership data and comprise the major input parameters to the first phase of the transportation demand analysis, i.e., trip generation. This information is available on a zonal basis and is discussed in the following section.

Transportation Demand Analysis

The analytical approach described in this chapter is the sequential transportation demand forecasting process. The main features of this process were described in Chapter 1 and are repeated in more detail in Figure 2-4. The process of calibrating and validating each of the demand forecasting models within the forecasting system is described in several operating manuals and texts. Normally, for the application of the models in preliminary or sketch planning, calibration is not conducted, and previously documented values of the input variables enable the estimates to be made with sufficient accuracy. Each of the phases of demand analysis is described in more detail in this section.

Trip Generation

Description

Trip generation is the process of estimating the number of trip ends for each of the study area zones. These estimates are based upon the measures of urban activity described in the preceding phase, related through mathematical relationships to the rate at which trips are made by various categories of trip makers.

Trip categories and purposes

Several basic geographic categories of trips comprise the total trips for most transportation studies:

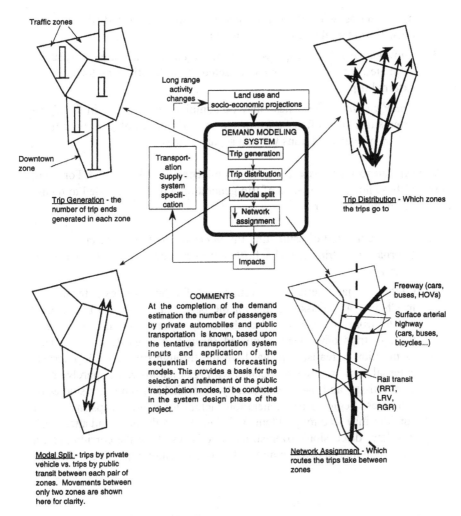

Traffic zones

Downtown zone

Long range activity changes

Land use and socio-economic projections

DEMAND MODELING SYSTEM

Trip generation

Trip distribution

Modal split

Network assignment

Transport- ation Supply - system specifi- cation

Impacts

COMMENTS
At the completion of the demand estimation the number of passengers by private automobiles and public transportation is known, based upon the tentative transportation system inputs and application of the sequential demand forecasting models. This provides a basis for the selection and refinement of the public transportation modes, to be conducted in the system design phase of the project.

<u>Trip Generation</u> - the number of trip ends generated in each zone

<u>Trip Distribution</u> - Which zones the trips go to

<u>Modal Split</u> - trips by private vehicle vs. trips by public transit between each pair of zones. Movements between only two zones are shown here for clarity.

<u>Network Assignment</u> - Which routes the trips take between zones

Freeway (cars, buses, HOVs)

Surface arterial highway (cars, buses, bicycles...)

Rail transit (RRT, LRV, RGR)

Figure 2-4 Demand forecasting process   *Source: Ref. (9)*

1. *Internal trips.* With both ends within the study area, these trips are usu-
    ally broken down into the trip purposes of home-based work (HBW)
    trips, home-based other (HBO), and non-home-based (NHB). Other
    subcategories may also be used. The information is usually based upon
    the results of home interviews of transportation users within the study
    area.

2. *Internal-external trips.* Because one end of the trip is in the study area
    and the other outside of it, information on these trips is usually based

upon data obtained at roadside interview stations located where routes cross the external study area cordon or boundary.

3. *Through trips.* These trips, with neither end in the study area, are usually estimated from the same data collected at the locations as for internal-external trips.

4. *Truck and taxi trips.* These may be combined with others in the above groups and may often be estimated as percentages of passenger trips, unless special conditions warrant separate surveys.

The trip categories above are illustrated conceptually in Figure 2-5. For the purposes of these notes, the process of trip generation will be simplified to focus on the key elements, as follows:

1. Only the method of estimating the *internal* trip ends is described here. Internal zone trips may be further broken down as "interzonal" and "intrazonal."

2. Of the internal-external and through trips, for the purposes of the project in Chapter 4, we assume that all the base year trips have been determined from surveys and that the design year internal-external and through trips have been estimated *through a separate process.* This assumption is realistic in that both the through trips and the internal-external trips are often associated with regional instead of purely urban area activity, and may consequently be available from other studies or estimates. Regional trip generation factors for internal trips in a small corridor may lead to: problems in balancing productions and attractions (see later discussion), overestimation of trips along the corridor, and a corresponding underestimation of out-of-corridor trips. These discrep-

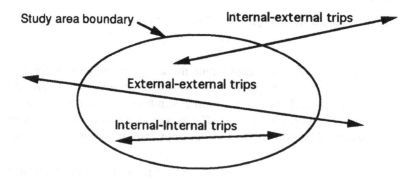

Figure 2-5 Trip categories

ancies may be adjusted by means of the screenline and cutline counts described later.

Trips may also be defined as person trips or vehicle trips. Most internal and internal-external trips are initially estimated as person trips, and through trips and truck and taxi trips are estimated as vehicle trips. However, the appropriate terminology, i.e., person or vehicle, should be checked in each case.

Analysis

Methods used to estimate trip generation include regression analysis, category analysis (sometimes referred to as cross-classification analysis), and trip rate analysis. Trips are usually estimated by category (HBW, HBO, NHB, etc). As described below, the most common methods used in most urban transportation planning studies are cross-classification analysis and trip rate analysis, the latter sometimes also incorporating regression analysis as a means of developing the trip rates.

Method

The trip generation process estimates the numbers of trips produced at, and attracted to, each zone, and usually requires a balancing process to ensure that the total productions are equal to the total attractions. The processes are described below.

Data

The data typically used for category analysis comprise the estimates of dwelling units (DUs) and income (for estimating trip productions) and number of employees (for estimating trip attractions) for each of the zones described in the preceding phase, "urban activity." For estimating trip attractions the trip rates based upon an employee or a floor area attraction rate for each zone are provided to permit direct computation of total person trip attractions per zone.

Trip productions

The trip production estimation method shown here is the cross-classification technique, based upon the relationships between household income and automobile ownership per dwelling unit, and the resulting trips per household by trip category. The relationships are documented for a wide range of populations, and enable data to be "borrowed" between cities of similar characteristics for preliminary planning. A selection is documented in (10). An example of the relationships and their use to estimate the number of trip productions for a single zone is provided in Figure 2-6. It shows how the relationships between income and automobile ownership for specific zones can be used with trip rates per household to estimate the total number of trips produced by a zone.

This application of the cross-classification method is sometimes known as the FHWA simplified trip generation procedure.

Trip attractions

Trip attractions are usually estimated by means of the trip rate method. This method employs known or estimated trips per unit of land use multiplied by the quantity of the units, such as gross floor area, to obtain the total zonal trip attractions. Examples of trip rates for various land uses are shown in the Appendices. An example of computing the total trip generation of a zone that comprises several different land uses is provided in Figure 2-7.

Multiple zones—estimating and balancing productions and attractions

Most zones in a study area will comprise more than one land use and, therefore, will generate both productions and attractions. This example also illustrates the process of balancing productions and attractions. The balancing is necessary because in any trip generation study the separately computed productions and attractions are rarely equal. It is usual to adjust the total home-based trip attractions to equal the total home-based trip productions, and then to adjust the non-home-based productions to equal the non-home-based attractions. When there is no breakdown available between home-based and non-home-based productions and attractions, it is acceptable to adjust the total attractions to equal the total productions, in recognition of the fact that trip productions tend to be more reliable, and then to adjust each of the zone attractions by the same adjustment ratio. Figure 2-8 provides an example of this balancing process. Figure 2-9 illustrates the trip generation process for a multizone area, based upon specified input data, and also incorporating the production-attraction balancing process.

Terminology for trip productions and attractions, and origins and destinations

The trip ends estimated in the trip generation phase represent trip productions and attractions and do not include the notion of trip directions. The terms "origins and destinations" imply this notion, and are usually employed at later stages of the analysis, such as following the trip distribution phase and before the modal split and traffic assignment phases. The differences are explained at that point, and are illustrated with an example.

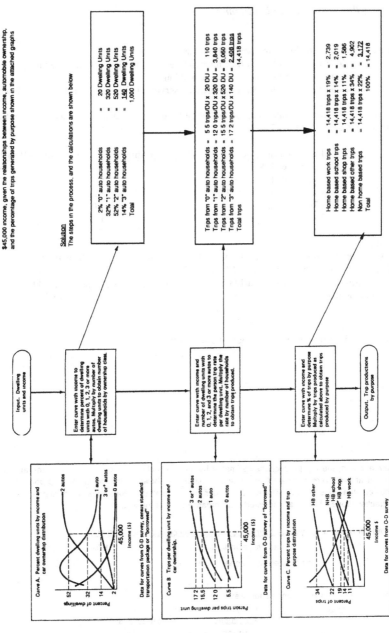

Figure 2-6 Trip production method and example *Source:* Based upon Ref. (1)

Three zones have the following predominant land-use areas

Zone 1	1 million sq ft commercial (retail), 2 million sq ft manufacturing
Zone 2	1.5 million sq ft public buildings
Zone 3	1 million sq ft commercial (service), 0.5 million sq ft commercial (wholesa

Given the trip attraction rates indicated in the table below, estimate the number of person trips attracted to zones 1, 2, and 3.

Land use Category	Daily Trip Attractions Per 1,000 sq ft [91 m²]
Commercial (Retail)	8.1
Commercial (service)	5.2
Commercial (wholesale)	1.2
Manufacturing	1.0
Public Buildings	1.5

Solution:
Zone 1 trip attractions = 1 million/1,000×8.1   + 2.0 million/1,000×1.0 = 10,100
Zone 2 trip attractions = 1.5 million/1,000 × 3.9   = 5,850
Zone 3 trip attractions = 1 million/1,000×5.2 + 0.5 million/1,000×1.2 = 5,800
Total attractions   21,750

Figure 2-7   Example of estimating trip attractions

For three zones, the trip productions estimated by the cross-classification method, and the attractions estimated above, were compared and found to differ, as shown in the columns for productions and unadjusted attractions, below.   Therefore, in order that the total attractions equal the total productions the trip attractions for each zone are adjusted in accordance with the formula:

Adjusted attr./zone = (unadjusted attr./zone) x (total area productions) = 23,800
                                        (total area attractions).   21,750

                              = unadjusted attr./zone x 1.09435

Estimate the adjusted attractions as follows:

	Productions	Unadjusted Attractions		Adjusted Attractions
Zone 1	11,000	10,100	(x 1.0934)	11,052
Zone 2	7,000	5,850	(x 1.0934)	6,401
Zone 3	5,800	5,800	(x 1.0934)	6,347
Total	23,800	21,750		23,800, checks with total productions.

Figure 2-8   Example of balancing trip productions and attractions

Problem

Estimate the trip productions and adjusted attractions for zones 1, 2, and 3 based upon the information provided as follows:

Zone	DUs	Avg Income Per DU	Employees	Trips per Employee
1	75	32	8	16
2	45	24	8	12
3	75	16	30	12

DUs, Income, and Employees, are in 1,000s

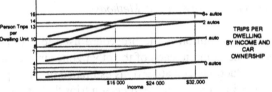

CROSS-CLASSIFICATION RELATIONSHIPS FOR HOUSEHOLD TRIP GENERATION ANALYSIS

PERCENT DWELLING UNITS BY INCOME AND CAR OWNERSHIP

TRIPS PER DWELLING BY INCOME AND CAR OWNERSHIP

Solution

The calculations are arranged in the table shown below. The FHWA Simplified Method Figure (2-6) is used for estimating the productions and trip rate analysis for estimating the attractions

For the productions, the percentages of DUs by auto ownership for the specified zonal incomes are read from the graph "percent of DUs" vs income and multiplied by DUs to give total DUs by autos owned. These are multiplied by Trips per DU, read from graph of person trips per DU to give Total Production

The attractions are calculated by multiplying the number of employees by the relevant trip rate. The total attractions are made identical to the total productions and the attractions for each zone adjusted by the ratio Total Unadjusted Attraction/Total Production

				Productions					Attractions			
Zone	DUs	Avg Income Per DU	Autos Per DU	Percent DUs by Autos Owned	Total DUs by Autos Owned	Trips Per DU	Total Products P[i]	Employees	Trips per Employee	Total Attracts	Adjusted Total Attracts A[i]
1	75	32	0	3	2.25	4	9.00	8	16	128	311.33
			1	42	31.50	10	315.00				
			2	45	33.75	14	472.50				
			3+	10	7.501	16	120.00				
			Zone Totals	100	75.00		916.50				
2	45	24	0	7	3.15	3	9.45	8	12	96	233.49
			1	52	23.40	8	187.20				
			2	35	15.75	13	204.75				
			3+	6	2.70	16	43.20				
			Zone Totals	100	45.00		444.60				
3	75	16	0	14	1.05	2	2.10	30	12	360	875.60
			1	60	4.50	7	31.50				
			2	21	1.58	13	20.48				
			3+	5	0.38	14	5.25				
			Zone Totals	100	7.50		59.33				
Totals							1,420.43			584	1,420.43

Note: DUs, Income, Employees, Total Productions, and Total Attractions in 1,000s

Figure 2-9 Example of trip generation—Multiple zones

Outputs

A list of the productions and attractions for each zone in the study area is the main output of this phase. As indicated earlier, the productions and attractions addressed in this section have included only those generated within the study area. Illustration of inclusion of the other internal-external trips and through trips will be made in the example shown in Chapter 4.

Linkages to Next Phase

The productions and attractions estimated in this phase provide the basic inputs to the next phase, trip distribution.

Trip Distribution

Description of Analysis Methods

Trip distribution is the procedure used to estimate the number of trips from each zone going to each of the other zones in the study area. The procedure uses the productions and attractions estimated in the trip generation phase and results in a trip interchange table of person trips that provides the basic inputs to the succeeding phase, modal split.

Of the three main methods of conducting trip distribution, i.e., the intervening opportunities model, the Fratar model, and the gravity model, the gravity model is the one most frequently used, and will be described here.

Data

The data for inputs to the gravity model trip distribution phase are as follows:

1. Trip production and attractions are those estimated in the preceding trip generation phase.

2. Travel times between zones are based upon estimated average peak hour travel speeds, which are representative of typical values experienced in similar cities or assumed for the functional classification of the highway and/or transit network being considered. These travel times are also similar to those used for the trip assignment stage described later.

3. Friction factors are determined from a graph of travel time versus friction factor. The friction factor is usually some form of the reciprocal of the travel time. It measures the impedance between zones and is calibrated for the study area prepared for the analysis, or may be obtained approximately from graphs for cities and transportation systems similar

to those in the study area. In certain cases, if significant trip costs and travel distances exist, these may be factored into the friction factor to provide a more realistic measure of the impedance perceived by travelers.

Basic gravity model theory

The gravity model is based upon similar concepts to the law of gravity, which states that the attraction between two bodies is proportional to the mass of each of the bodies and inversely proportional to the distance between them. The gravity model's formulation of this concept relates the number of trips between each pair of zones to the attractions and an impedance between each pair of zones to the sum of the attractions and impedances for all the zones in the system. This formulation and an approach useful in conducting the necessary balancing between attractions following each iteration of the calculations are summarized in Figure 2-10.

Gravity model calculations

Two illustrations of the application of the gravity model follow. The first, based on (12) is shown in Figure 2-11. It depicts the case of estimating the number of trips from a single zone to several other zones, where no balancing of attractions is required. The second, in Figure 2-12, illustrates the case where three or more zones experience productions and attractions in each zone and the number of trips to and from each zone is to be estimated. The method of conducting the computations is that used in addressing "quick response" techniques in urban travel estimation, detailed in (11). The layout adopted in this method enables the computations to be done manually, and may be readily adapted for use with computerized spreadsheets.

In addition, the procedure for converting the matrix of trip interchanges for the 24-hour trips to a triangular matrix of trips between the zones (i.e., the sum of both directions) is shown. This matrix may then be multiplied by the ratio of design hour to daily trips (usually approximately equivalent to the ratio Design Hourly Volume/Average Daily Traffic (DHV/ADT) and known as the **K** factor (not to be confused with the socioeconomic adjustment factor used in the gravity model itself) to provide the number of trips during the design hour between zones for direct input to the next phase, modal split. The value of **K** is typically used in traffic studies, and its use here assumes that its value is applicable to transit users as well as automobile users. Also, it should be noted that the conversion to a design hour basis may be done after the modal split has been completed, but for the purposes described in this book, i.e., design of a system based upon the design hour demand, the conversion to peak hour values simplifies the calcula-

The Gravity Model is mathematically expressed as:

$$T_{ij} = P_i \frac{A_j F_{ij} K_{ij}}{\sum_{j=1}^{n} A_j F_{ij} K_{ij}} \tag{1}$$

where

and

$$F_{ij} = f(tij)$$

T_{ij} = trips produced in analysis area i, and attracted at analysis area j;

P_i = total trip production at i;

A_j = total trip production at j;

F_{ij} = friction factor for trip interchange ij;

K_{ij} = socioeconomic adjustment factor for interchange ij if necessary;

tij = travel time (or impedance) for interchange ij;

i = origin analysis area number, $i = 1, 2, 3, \ldots, n$;

j = destination analysis area number, $j = 1, 2, 3, \ldots, n$;

n = number of analysis areas

For the manual application, Kij has been discarded altogether. The Gravity Model formulation then simplifies to the following form for manual application:

where

$$T_{ij} = R_i A_j F_{ij}$$

$$R_i = \frac{P_i}{\sum_{j=1}^{n} A_j F_{ij}} \tag{2}$$

called the "production index" (a constant for each production analysis area i)

$A_j F_{ij}$ = the "attraction factor" for analysis area j and;

$\sum_{j=1}^{n} A_j F_{ij}$ = the "accessibility index" for analysis area i.

Mathematically, the Gravity Model is formulated so that a production balance is maintained; in other words, the produc-

tion (row) totals for each analysis area as calculated from the model equal the input productions. However, the attraction (column) totals for each analysis area output from the model will not necessarily match the desired input values. To attain an acceptable attraction balance, an interative process is employed to adjust the calculated trip interchanges.

After each application (iteration) of the Gravity Model, the adjusted attraction totals (for each analysis area) to be used for the next iteration are calculated according to the following formula:

$$A_j^q = A_j^{q-1} \cdot \frac{A_j}{C_j^{q-1}} \tag{3}$$

where

A_j^q = adjusted attraction factor for attraction analysis area (column) j, iteration q; for manual application only a maximum of one additional iteration is suggested;

$A_j^{q-1} = A_j$, when $q = 1$

C_j^{q-1} = attraction (column) total for analysis area j, resulting from the application of the Gravity Model during iteration $q - 1$;

A_j = Original and desired attraction total for attraction analysis area (column) j; this is the value developed from the trip generation step;

j = attraction analysis area, $j = 1, 2, \ldots, n$;

n = number of analysis areas;

q = iteration number, $q = 1, 2, \ldots, m$.

For the manual trip-distribution procedure, two iterations are considered sufficient. For many uses, only the original application (the first iteration) with no adjustment could be considered adequate. The option of whether or not to iterate twice depends on the percent difference between the attraction totals at the end of each iteration and that originally input for each analysis area. Generally, a 5- to 10-percent difference is acceptable, depending on the degree of accuracy required.

Figure 2-10 Trip distribution using gravity model—Theory *Source*: Ref. (11)

tions and is more consistent with the quantification of the policy variables that apply during the peak hour travel period. It should be noted that the entire trip generation and distribution process may be done for the peak hour only, but this is not usually the case for long-range studies.

Outputs

The outputs of this phase are a trip table that shows the number of trips produced in each zone and attracted to each other zone in the study area for the 24-hour period, then the 24-hour trips between the zones (triangular matrix) and, following conversion by a factor **K,** the trips between each pair of zones during the design hour.

The results of the trip distribution, known as the trip interchange table, using the values computed in the design hour, are summarized in Figure 2-13. They

The Gravity Model is mathematically expressed as:

$$T_{ij} = P_i \ \frac{A_j F_{ij} K_{ij}}{\sum_{j=1}^{n} A_j F_{ij} K_{ij}}$$

where

$$F_{ij} = f(tij)$$

and

T_{ij} = trips produced in analysis area i, and attracted at analysis area j;
P_i = total trip production at i;
A_j = total trip attraction at j;

F_{ij} = friction factor for trip interchange ij;
K_{ij} = socioeconomic adjustment factor for interchange ij if necessary;
tij = travel time (or impedance) for interchange ij;
i = origin analysis area number, $i = 1, 2, 3, \ldots, n$;
j = destination analysis area number, $j = 1, 2, 3, \ldots, n$;
n = number of analysis areas.

For the manual application, K_{ij} has been discarded altogether. The Gravity Model formulation then simplifies to the following form for manual application:

$$T_{ij} = P_i \ \frac{A_j F_{ij}}{\sum_{j=1}^{n} A_j F_{ij}}$$

EXAMPLE - DISTRIBUTION FROM A SINGLE ZONE

Problem:

Estimate the number of trips distributed from Zone 1 to each of Zones 2, 3, and 4, given the productions, attractions, and travel times between the zones and the associated travel time factors shown in the diagram, graph, and table below.

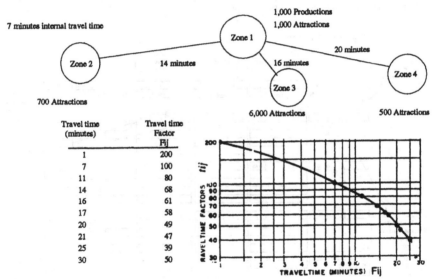

Travel time (minutes)	Travel time Factor Fij
1	200
7	100
11	80
14	68
16	61
17	58
20	49
21	47
25	39
30	50

Solution:

One way of conveniently calculating the number of trips is to arrange the calculations as shown in the table below. The solution to the problem is shown in the last column of the table. The trips, Tij, from Zone 1 to each of the zones, are obtained by application of the equation presented above.

Zone	Aj	tij	Fij	AjFij	Tij
1	100	7	100	1,000	186
2	700	14	68	476	88
3	6,000	16	61	3,660	680
4	500	20	49	245	46
			Sum, AjFij =	5,381	
			Sum Tij =		1,000

Figure 2-11 Trip distribution—Theory and basic example

Problem

Three zones in an urban area are characterized by the following trip productions and attractions (derived from a trip generation analysis), travel times between zones (tij) and associated friction factors (Fij). Conduct a gravity model trip distribution in order to prepare a 24-hour trip interchange table and triangular matrix, and a desire line diagram. In this problem, zone 1 is CBD, Zone 2 is a central city area and Zone 3 is predominantly a residential area. The total number of trips in the 24-hour period is 100,000.

In this problem there is no breakdown between trip purposes and it is assumed that the friction factors associated with each of the travel times between zones represent a consolidated value.

Zone	Pi	Aj
1	500	6,000
2	3,500	1,000
3	6,000	3,000
Total	10,000	10,000

Productions and attractions (in 1,000s)

		Attractions 1	2	3
	1	17.0	17.4	15.9
Prods	2	17 4	6.0	8.3
	3	15.9	8.3	8.8

Travel times (tij) between zones, min.

		Attractions 1	2	3
	1	0.65	0.60	0.80
Prods	2	0.60	6.00	3.30
	3	0.80	3.30	3.00

Friction factors (Fij) between zones

Solution

A useful format for assembling the input data and conducting the analyses for two iterations of the gravity model is described in the five steps below and the remaining steps on the next page.

24-Hour Passenger Trips in 1,000s

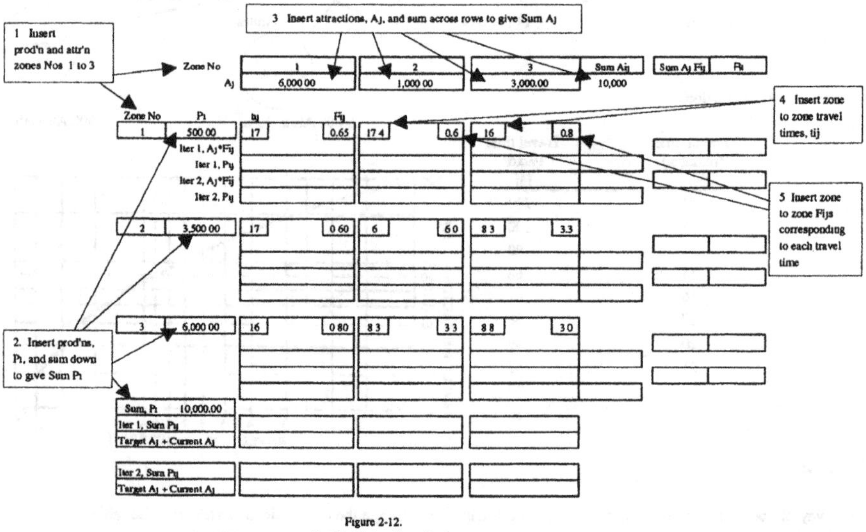

Figure 2-12.
Trip Distribution Example

Figure 2-12 Trip distribution example

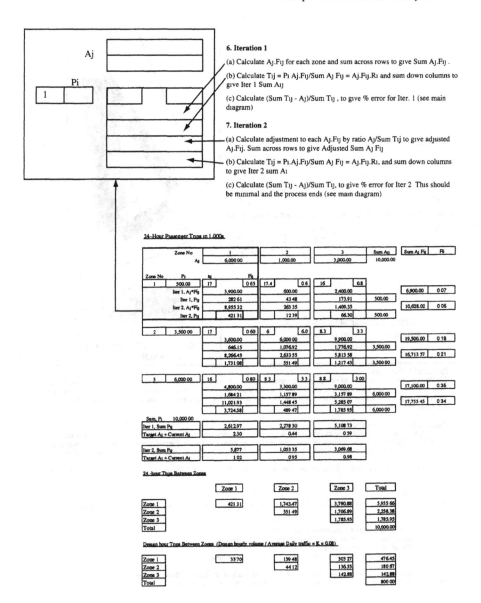

6. Iteration 1

(a) Calculate Aɉ.Fɪɟ for each zone and sum across rows to give Sum Aɉ.Fɪɟ .

(b) Calculate Tɪj = Pɪ Aɉ.Fɪɟ/Sum Aɉ Fɪɟ = Aɉ.Fɪɟ.Rɪ and sum down columns to give Iter 1 Sum Aɪɟ

(c) Calculate (Sum Tɪɟ - Aɉ)/Sum Tɪɟ , to give % error for Iter. 1 (see main diagram)

7. Iteration 2

(a) Calculate adjustment to each Aɉ.Fɪɟ by ratio Aɉ/Sum Tɪj to give adjusted Aɉ.Fɪj. Sum across rows to give Adjusted Sum Aɉ Fɪɟ

(b) Calculate Tɪj = Pɪ.Aɉ.Fɪɟ/Sum Aɉ Fɪɟ = Aɉ.Fɪɟ.Rɪ, and sum down columns to give Iter 2 sum Aɪ

(c) Calculate (Sum Tɪɟ - Aɉ)/Sum Tɪɟ, to give % error for Iter 2 This should be minimal and the process ends (see main diagram)

Figure 2-12   continued

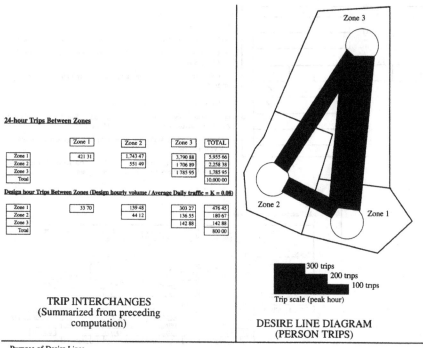

24-hour Trips Between Zones

	Zone 1	Zone 2	Zone 3	TOTAL
Zone 1		421 31	1,743 47	3,790 88	5,955 66
Zone 2			551 49	1 706 89	2,258 38
Zone 3				1 785 95	1,785 95
Total					10,000 00

Design hour Trips Between Zones (Design hourly volume / Average Daily traffic = K = 0.08)

	Zone 1	Zone 2	Zone 3	Total
Zone 1		33 70	139 48	303 27	476 45
Zone 2			44 12	136 55	180 67
Zone 3				142 88	142 88
Total					800 00

TRIP INTERCHANGES
(Summarized from preceding
computation)

300 trips
200 trips
100 trips
Trip scale (peak hour)

DESIRE LINE DIAGRAM
(PERSON TRIPS)

Purpose of Desire Lines

The main purpose of the desire line diagram is to present a visual indication of the extent and pattern of the reported (or future projections of the) spatial distribution of trips throughout the study area which are desired by trip makers. It should be noted that the desire line diagram for a base year affects the travel desires of the trip makers given the conditions of the transportation facilities and land uses at that time

Preparing the Desire Line Diagram

The number of trips from each zone to each of the other zones is summed graphically. Probably the simplest way of doing this is to first draw a single line between each of the zone pairs and list the volume alongside. Second, select a volume scale which enables the greatest sum of the trips (usually occurring adjacent to the CBD) and draw each desire line to scale between the appropriate zones

It also may be useful to prepare a desire line diagram or person trips by private automobile and also for transit trips, and for the total together.

Interpreting the diagram

The desire line diagram assists in conducting a quick check of trip distribution, and typically emphasizes the following for an urban area:

- A preponderance of trips oriented to the CBD.
- Large numbers of trips which focus on other major generators.
- The presence of through trips
- Cross-town trips generally distributed throughout the study area.

Usually, a desire line diagram is prepared for a 24-hour period of a typical weekday. However, peak hour or other periods, cargo movements only, bicycle movements or other specific modes can also be shown if necessary.

Figure 2-13   Trip interchange summary and desire line diagram

represent the total person trips between zones during the peak hour. Also shown is the desire line diagram, which summarizes the trip interchanges graphically. A brief explanation of how the desire line diagram is constructed and interpreted is also given.

Defining Trips Productions and Attractions, and Origins and Destinations

Generally, trips based upon household characteristics are the easiest to identify and forecast. For this reason, the home is often used as a base to predict travel. This convention requires a distinction in the terms *production* and *attraction,* and *origin* and *destination.* The production and attraction trip ends estimated in the trip generation phase do not imply any notion of direction. Therefore, the term "origin" is used to denote the beginning of a trip and the term "destination" to denote the termination of a trip. The rules for their definition are listed below (1). An example illustrating the conversion from productions-attractions to origins-destinations, and construction of a triangular matrix (the number of trips between zones) for a daily (24-hour) and a peak hour period are shown in Figure 2-14.

Rule 1. Trips that either begin or end at the traveler's home are *produced* at the *home* end.

Rule 2. Trips that either begin or end at the traveler's home are *attracted* to the *non-home* end.

Rule 3. Trips that begin at a non-home location and end at another non-home location are *produced* at the origin and *attracted* to the destination.

For example, a traveler goes from home to the office, to a store, then back home:

1. *Trip number 1* was produced at home and attracted to the office—Rules 1 and 2—therefore a home-based work (HBW) trip.

2. *Trip number 2* was produced at the office and attracted to the store—Rule 3—therefore a non-home-based other (NHBO) trip.

3. *Trip number 3* was produced at home and attracted to the store—Rules 1 and 2—therefore a home-based other (HBO) trip.

Gravity Model Approximations

In presenting the foregoing example of the gravity model estimation procedure, certain assumptions and approximations have been made in accordance with the objective of illustrating the major features of the process and the sketch planning level of detail. They include the following:

- No breakdown for trip purpose is made. Typically, the gravity model would be applied for each of the major trip purposes (home-based work, home-based nonwork, etc.) with associated friction factors. The composite gravity model used here, therefore, represents all trip purposes and the friction factors related to travel speeds were developed judgmentally to represent them.

Problem

The three zones and trips shown on the right illustrate the trip movements in a given time period. Based upon the definitions given, construct three tables, or matrices, showing origins (O), destinations (D), productions (P), and attractions (A), and the number of trips between zones (a triangular matrix).

Solution

1. List the Os, Ds, Ps, and As for each of the zones, and note at which end of the trip these occur. This is an intermediate step, and helps to clarify the process.

2. Construct the three matrices, i.e. the O-D, P-A, and the triangular, using the values shown in the diagram on the right. The three matrices are shown below.

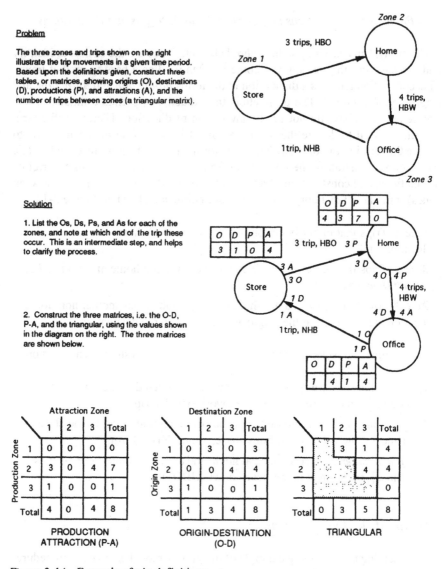

Attraction Zone

Production Zone	1	2	3	Total
1	0	0	0	0
2	3	0	4	7
3	1	0	0	1
Total	4	0	4	8

PRODUCTION
ATTRACTION (P-A)

Destination Zone

Origin Zone	1	2	3	Total
1	0	3	0	3
2	0	0	4	4
3	1	0	0	1
Total	1	3	4	8

ORIGIN-DESTINATION
(O-D)

	1	2	3	Total
1		3	1	4
2			4	4
3				0
Total	0	3	5	8

TRIANGULAR

Figure 2-14 Example of trip definitions

- No socioeconomic adjustment factor K is applied. This factor would normally be developed during the calibration of the model and, in some cases, may not be required.

- In many cases, the 24-hour trip distribution is not converted to a peak hour distribution until the modal split or traffic assignment phase. However, because of the more limited objectives of preliminary studies, and the reduced computational effort of using the design hour instead of 24-hour volumes for the following phase, modal split, the peak hour movements are used in this case.

Alternative Method of Computing Peak Hour Trip Interchanges

Instead of using a gravity model friction factor for all trips, the factor may be obtained for HBW trips only and then a second factor applied to convert the work trips to design hour trips. Graphs and tables for doing this are provided in (11). The trip generation phase will also have to be modified accordingly to provide the HBW trip interchanges. The total difference in computational effort is not likely to be significant between the two methods, although it is more usual to find the friction factors documented by trip purpose than for all trips.

Modal Split

Description

The process of determining which mode of travel (automobile or public transportation) each person will use between the origins and destinations determined in the preceding trip distribution phase is called modal split, or modal choice, analysis. Also in this phase, person trips for the peak hour period are converted to vehicle trips, by means of an automobile occupancy factor, and to transit passenger trips.

The modal split phase is the first in the demand analysis process in which several policy variables may be introduced; specifically, the travel time and cost components. This is detailed in the description below, based upon (1, 13), and other references.

Method

Several methods of conducting modal split analysis have been developed and used. They include regression analysis, stratified diversion curves, and the use of logit analysis based upon the perceived utility of specific modes by users. Other methods, used for approximate analyses, may include elasticity models and historical documentation of modal change related to variables such as travel time and travel cost. Elements of these methods are described in Chapter 5.

Logit analysis

The method of conducting modal split analysis that has gained most popularity in recent years is the logit model method. It provides an estimate of the proportion of persons using public transportation between any two zones, based upon the perceived utilities of each mode by users in each zone. The basic approach and an example are shown in Figures 2-15 and 2-16, respectively. The example shows the necessary input values used in the utility equations and a tabular method of assembling the data and carrying out the computations.

Trip Interchange Mode Usage Models

As shown in the diagram below, the trip interchange models are applied after the trip distribution phase, i.e. the trip usage analysis is based upon the interchange of trips between the pairs of origins and destinations.

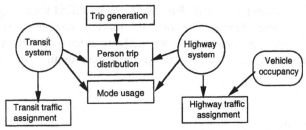

The Logit Model of Trip Usage

The logit model is a share model that divides the trips between the various modes depending on each mode's relative desirability for any given trip. A mode is relatively more desirable (has greater utility) if it is faster, cheaper, or has other more favorable attributes than competing modes. The utility of a mode may be expressed by means of a linear utility expression such as the following for the automobile mode:

$U_{Auto} = 0.25 + 1.00$ (in vehicle time) $+ 2.5$ (out of vehicle time) $+ 0.33$ (out-of-pocket cost)

Each variable represents some characteristic of the alternative which helps to distinguish it from the others. The relative influence of each variable in determining the desirability of the alternative is given by the weight coefficient.

The logit model takes the following form to trade off the relative utilities of the various modes:

$$\text{Probability of using mode i} = \frac{e^{(\text{utility of mode i})}}{\displaystyle\sum_{n=1}^{\text{all modes}} e^{(\text{utility of mode n})}}$$

The exponential function "e" gives the characteristic logit curve shown in the diagram below for the two mode case. Additional modes can be considered but cannot be represented by a two-dimensional plot as shown.

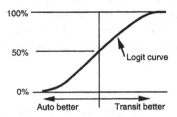

Figure 2-15 Trip interchange and logit model modal split—Theory

Problem:

The results of a trip distribution between two zones show that 600 trips will occur on a 24-hour basis. Estimate the modal split of these trips between automobile and transit, and the number of automobiles and transit passengers if the automobile occupancy is 1.2 persons per vehicle.

Assume utility of mode K, U(K): $U(K) = a(K)-0.07S-0.015L-0.0025C,$ where,

a(K)=Calibrated constant for specified mode K, (-0.8 for auto, -0.1 for transit)
S=Service time (walking and waiting) in minutes
L=Line-haul time in minutes
C=Out-of-pocket costs in cents

Solution:

Using the multinomial logit model described on the preceding page, the design year service attributes may be listed and the modal split p(K), calculated as follows:

Mode	Attribute				Utility	Exponent	Modal Split
	a	S	L	C	U(K)		p(K)
Auto	-0.80	5.00	28.00	580.00	-3.02	0.05	0.34
Transit	-0.10	20.00	42.00	85.00	-2.34	0.10	0.66
						Total........	1.00

Transit Person and Auto Vehicle Trips

Person trips from Trip Distribution phase (given)	600
Percentage transit trips (from MS, above)	0.66
Total transit person trips	396
Auto person trips (total trips - transit person trips)	204
Ave auto occupancy	1.20
Auto vehicle trips	170

Figure 2-16   Example of modal split using multinomial logit model

It should be noted that household auto availability and age of traveler on transit use may affect modal split considerably. Therefore, if "captive" transit users comprise a large portion of the total travelers—particularly in zones with large numbers of school children or elderly people—a trip generation model based upon the person and housing unit characteristics of the traveler, as opposed to mainly service characteristics, may be a better predictor of mode than a logit model, particularly for work trips.

Data

The major data inputs required to conduct the modal split analysis using the logit model are:

1. The number of design hour, person trip interchanges between zones resulting from the trip distribution phase of the project.

2. Estimated service time (walking and waiting) between zones for automobile and public transportation modes. These would include the cost of tolls, parking, and gasoline for automobiles, and transit fares for public transportation users.

3. The vehicle occupancy values used to convert person trips to vehicle trips.

4. A list of the pairs of zones between which public transportation service is available. It is assumed that all zones will be accessible between each other by automobile, but connections between some zones may not be possible by public transportation.

5. The utility equation constants based upon surveys of travel mode choice or experience gained in cities with similar demographic and transportation facilities.

Person trips versus vehicle trips

The person trips by private vehicle are converted to vehicle trips by dividing person trips by the vehicle occupancy factor. Person trips by public transport are left as person trips for input to the next phase—traffic assignment. They are converted to transit vehicle volumes in the facilities design phase, described later, because until the link volumes are known, the appropriate transit mode and, hence, the vehicle occupancy, cannot be determined.

Outputs

A matrix showing the trip interchanges from the trip distribution analysis, broken down into automobile users and public transportation users, is the major output of this phase.

Approximations

Normally, as with trip generation and trip distribution, the modal split procedures described here would be conducted for each of the various trip purposes. Instead, for simplicity, the utility expressions detailed here represent a judgmental composite of all daily trips for all purposes for peak hour travel, based upon examples for existing cities.

Total travel times for each of the modes are approximate because, until the actual modal split and the type and extent of the facilities are known, the associated travel speeds are also only approximate. This problem is also encountered with more detailed models of demand and, ideally, the process requires several iterations before a reasonably satisfactory representation of the travel times is attained, but this is often not done in practice.

Traffic Assignment

Description

This phase predicts the paths that the vehicles and transit passengers will take on the links of the corridor between each of the zones, and assigns the appropriate volumes to each of these links by an additive procedure. The basic method of conducting the trip assignment is by means of the "all or nothing" method, which assigns all traffic to the least time route between zone centroids as initially determined, with no modifications for changes in travel speeds due to congestion as the volume of traffic on any particular link increases. This method, described in (1, 14), and other references, is used in these notes. Refinements such as the probabilistic routing of traffic or the capacity restraint method are not discussed here.

Although there are differences in conducting the assignment for highway versus transit trips, the principles remain the same, and the example of the highway trip assignment described below could apply to the transit assignment in its essential elements.

Data

The following data are required:

1. A map showing the possible routes and the corridor and the associated link and node map with initial estimates of travel times for each link.
2. The private vehicle and transit passenger trip interchanges between the zones of the study area. These data are obtained from the foregoing modal split phase.

Method

The steps in conducting the trip assignment process are as follows:

1. Network development
2. Network coding
3. Network loading
4. Network calibration.

The above steps are illustrated in the simplified example described below and shown in Figure 2-17. This example considers automobile trips on a highway network.

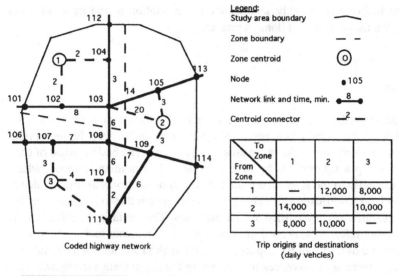

Coded highway network

Legend:
Study area boundary	—⌒—
Zone boundary	— —
Zone centroid	⊚
Node	•105
Network link and time, min.	•—8—•
Centroid connector	—2—

Trip origins and destinations
(daily vehcles)

To Zone / From Zone	1	2	3
1	—	12,000	8,000
2	14,000	—	10,000
3	8,000	10,000	—

Minimum time path

From zone	To zone	Minimum Time Path	Time (minutes)
1	2	1-104-103-108-109-2	21
	3	1-104-103-108-110-3	20
2	1	2-109-108-103-104-1	21
	3	2-109-111-3	10
3	1	3-111-110-108-103-104-1	20
	2	3-111-109-2	10

Assigned vehicle volumes, link 103-108

Zone Pair	Vehicle Volume	
	Northbound	Southbound
1-2	—	12,000
1-3	—	8,000
2-1	14,000	—
2-3	—	—
3-1	8,000	—
3-2	—	—
Total	22,000	20,000

Problem
The coded highway network shown accommodates the vehicle trip origins and destinations for zones 1, 2, and 3, as indicated in the top two diagrams. Estimate the nortbound and southbound vehicle volumes on link 103-108, using the 'all or nothing' least time travel route assignment method.
Solution
The skim tree table is constructed to list, in order, the nodes on the least time routes between each of the zone centroids. The time for each of the least time routes is noted. Then, a table of the assigned volumes is constructed for each of the pairs of zones, and summed to provide the total vehicles in each direction for the required link.

Figure 2-17 Trip assignment example   *Source*: Ref. (9)

Network development

The highway network is shown in abstract form with the following features:

1. An abstract of the arterial and other major highway networks links and nodes, showing the travel time and distance in miles for each link. Travel time may be estimated from current or anticipated conditions for the initial assignment process. These times are also at the trip distribution stage to provide estimates of times between zones for the gravity model.
2. The location of the traffic zones, each of which is represented by a "centroid" defining the estimated center of trip attraction and production activity.
3. Centroid connecters linking the centroids to the highway network of interest. These connectors represent the local street system. The travel time related to the time of day for which the assignment is being performed for each connector is also shown. Judgement is required about which street to include in the abstract network, and this must be guided by the ultimate purpose of the particular assignment. Normally, local and minor streets are excluded from most areawide assignment studies.

Network coding

The centroids and the nodes are numbered so that each network link or centroid connector is defined by a pair of numbers.

Network loading using the minimum path technique

The network is "loaded" with the trips shown in the trip interchange table using the minimum path technique. To do this, "trees" from each zone to each other zone are defined. A tree is constructed by computing the least time path from each zone centroid to each other zone centroid. The tree is built successively using the terminal node nearest the origin of the trees as the next branching node. All links connected to the branching node are added to the tree and their terminal nodes become branching nodes. The shortest route between two specific centroids can thus be expressed in terms of the total travel time for the selected route.

The traffic flow on each highway link computed as shown for this example is for link no. 103-108 only. The example shows a mechanism for adding the traffic volumes that occur on link 103–108 from each skim tree, indicating a total of inbound movements and outbound movements. The same procedure would be carried out for all other links in order to complete the assignment for the entire network.

Network calibration

Although it is not addressed in these notes, calibration is done to ensure that the base year estimated traffic volumes correspond reasonably with the actual volumes that occur on the network (determined from counts). The differences in comparisons may be due to inaccuracies in the trips used for assignment, in the zone-network compatibility established, in the coding of geometrics and operational characteristics (i.e., travel time), or in the counts being used for comparative purposes. Checking of assignment volumes with actual vehicle counts is usually done in the following ways:

1. Comparing vehicle miles traveled (VMT) and vehicle hours of travel (VHT) from the assignment with ground counts over the entire study area and in subdivisions of the study area (i.e., corridors, rings, zones or districts) by functional class of highway.

2. Comparing actual volume counts with assigned volumes at synthetic screenlines. This may be done across long gridlines traversing the entire area for trip data checking, and across short cutlines for checking the assignment of traffic along corridors of travel.

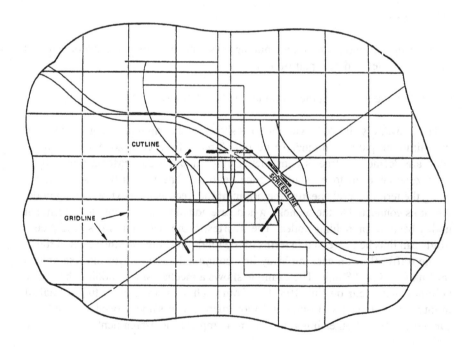

Figure 2-18   Locations of screenlines and cutlines   Source: Ref. (1)

3. Comparing volumes on specific links—usually those with high traffic volumes.

4. Root-mean-square error comparisons by volume group.

Examples of the location of screenlines and cutlines are shown in Figure 2-18. The estimated versus the actual traffic volumes on the corridor and/or road network links at these lines would be compared. If there was an unusually large difference between them (see below) the analysis process would have to be reviewed to discover possible errors and to correct them.

Levels of Accuracy in Trip Assignment

Specific levels of accuracy for use in all circumstances are difficult to prescribe because of the wide variety of highway functional classifications, traffic compositions, adjacent facilities, and quality of basic data available to the analyst. However, an accuracy that would not result in a difference of one lane in the design of a facility could be considered reasonable. On this basis, Figure 2-19 indicates the required level of accuracy for a given average daily traffic (ADT) volume together with examples of the accuracy of selected assignment studies conducted in the past (14, 15).

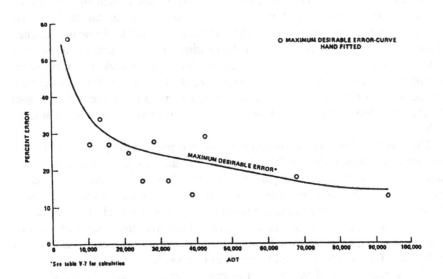

Figure 2-19 Percent error in assignment results related to average daily traffic (ADT)
Source: Refs. (14) and (15)

Outputs

The outputs of this phase are automobile volumes and transit passenger design hour volumes by predominant direction for each link in the network. These volumes are then used directly in the planning and design of the physical facilities and operational practices such as transit scheduling, described in Chapter 3.

Notes and Comments

In this example, the number of zones and the extent of the transportation network have been kept to a minimum in order to simplify the computations and make them amenable to the use of a pocket calculator. If the travel time on any link were to be altered to the extent that a different route between two nodes occurred, the table would have to be rearranged accordingly.

Although the examples have shown the results of the numerical analyses, the solutions should always be checked to see if volumes are reasonable with regard to the approximate capacity of each link in the network. Links with especially high or low volumes are particularly important. Comparisons with past examples are also valuable to ensure that the results are reasonable.

Review and Comments

This chapter has outlined the main features of the traditional demand forecasting process, with examples illustrating the application of typical data and analysis methods—several of them in a calculation format that will be found useful when conducting the project in Chapter 4. Estimation of the demand for transportation, defined by the volumes of vehicles on a highway network link or the number of passengers on a transit link, will provide the basic inputs for consideration of the modes of transportation that will best accommodate these volumes, also taking into account the physical and operational constraints that may exist in an urban area.

The phase of demand forecasting that has incorporated the policy variables to the greatest extent is the modal split phase. During this phase, the policy variables of service time for automobile versus transit users, and fare level and in-vehicle travel time for transit users, are included as attributes in the multinomial logit model. As an extension to the modal split phase, the level of vehicle occupancy is also a policy variable to the extent that governmental incentives for car- and vanpooling, such as provision of priority treatments at ramps, matching of riders, and fee reductions at toll booths, can be provided.

Ideally, the results of the demand estimation process should be re-estimated following the complete cycle of the facilities design process. This is because values of inputs assumed initially, such as travel time in the trip distribution phase, may vary depending on the characteristics of the mode selected, the level of service (LOS), and other physical and operational features. However, this re-estimation is often not done in practice at the sketch planning level of detail, and may be more appropriately completed during the refinement stage of a selected plan.

The outputs from demand forecasting (i.e., the volumes of highway vehicles and transit passengers on specific links of the system) will be needed as a basis for examining and selecting the preferred modes of transportation that will make up the physical and operating system. This is described in the next chapter.

References

1. U.S. Department of Transportation, Federal Highway Administration, Urban Mass Transportation Administration, *Urban Travel Demand Forecasting—A Self Instructional Text,* Washington, D.C., 1977.

2. Manheim, Marvin L., *Fundamentals of Transportation Systems Analysis,* vol. 1, MIT Press, Cambridge, MA, 1979.

3. Massachusetts Department of Public Works, *Central Artery/Third Harbor Tunnel Project, Draft Report, Detailed Travel Model Documentation,* Boston, MA, February 1990.

4. U.S. Bureau of the Census, *U.S. Bureau of the Census Tracts, Boston Metropolitan Area,* U.S. Bureau of the Census, Washington, D.C., 1990.

5. Wilbur Smith and Associates for the Automobile Manufacturers Association, *Future Highways and Urban Growth,* New Haven, CT, 1961.

6. Wilbur Smith and Associates for the Automobile Manufacturers Association, *Transportation and Parking for Tomorrow's Cities,* New Haven, CT, 1966.

7. Hanson, Susan. *The Social Geography of Urban Transportation.* The Guilford Press, New York, 1986.

8. Bhatt, Kiran *Potentials of Congestion Pricing in the Metropolitan Washington Region,* Special Report 242 Transportation Research Board, Washington, D.C., 1994.

9. Schoon, John G., *Transportation Systems Planning, Project Notes,* Department of Civil Engineering, Northeastern University, Boston, MA, 1993.

10. U.S. Department of Transportation, Federal Highway Administration, Urban Planning Division, *Trip Generation Analysis,* Washington, D.C., 1975.

11. National Cooperative Highway Research Program, Report No. 187, *Quick-Response Urban Travel Estimation Techniques and Transferable Parameters—Users Guide,* National Research Council, Washington, D.C., 1978.

12. U.S. Department of Transportation, Federal Highway Administration, *Calibrating and Testing a Gravity Model for any Size Urban Area,* Washington, D.C., reprinted 1973.

13. U.S. Department of Transportation, Bureau of Public Roads, *Modal Split*, U.S. Government Printing Office, Washington, D.C., 1966.[1]

14. U.S. Department of Transportation, Federal Highway Administration, *Traffic Assignment*, Washington, D.C., 1973.[1]

15. U.S. Department of Transportation, Federal Highway Administration, *Computer Program for Transportation Planning, PLANPAC/BACKPAC, General Information Manual*, Washington, D.C., 1977.

Other relevant texts and documents are listed in the Bibliography.

Selected Bibliography

U.S. Department of Transportation, Office of Transportation Economic Analysis, *Evaluating Transportation System Alternatives*, Washington, D.C., 1978.

U.S. Department of Transportation, Urban Mass Transportation Administration, *Characteristics of Urban Transportation Systems*, Washington, D.C., June 1979.[1]

U.S. Department of Transportation, Urban Mass Transportation Administration, *Transit Corridor Analysis—A Manual Sketch Planning Technique*, Washington, D.C., 1979.

U.S. Department of Transportation, Urban Mass Transportation Administration, *Characteristics of Urban Transportation Systems*, Washington, D.C., October 1984.[1]

[1]Excerpts from these publications are presented at various places in the text, figures, and tables of this book.

3

Facilities Design and Impacts

Introduction

The design of transportation facilities directly affects the levels of benefits and impacts that may be expected as a result of specific service policy variables. Therefore, design and the resulting impacts are considered together in this chapter and would normally follow the demand estimation processes described in Chapter 2. Again, the methods selected are presented in a form that is suitable for performing the estimations manually or by means of computer spreadsheets. The major elements are as follows:

Physical Facilities Design

- Location and physical and geographical constraints
- Capacity and level of service
- Available modes
- Mode combination.

Impact Estimation

- Capital cost
- Energy use
- Air pollution.

In this phase the overall volumes of automobile and transit passengers estimated in the demand analysis (traffic assignment phase) become translated into the supply of transport in terms of volumes of specific vehicle types such as automobiles, buses, and rail transit on the critical links of the network. This, in turn, enables the numbers of automobile and bus lanes and rail tracks to be estimated.

The design considers the modes available, their operating characteristics such as capacity and level of service (LOS), and the mix of modes that offers a balanced system. A summary of selected public transportation mode characteristics was included in Chapter 1, and further details are provided in the Appendices. The travel times and costs used earlier in the modal split analysis imply the use of certain travel speeds and LOS. These have a direct effect on the planner's selection of the modes used, their operating characteristics, attainment of the policy goals described earlier, and the extent of the various impacts.

Physical Facilities Design

Location of Transportation Facilities in Urban Areas

The physical locations of the highway lanes and/or rail tracks within the urban area are affected by the proximity of existing transportation facilities and residential, commercial, and recreational land uses. Other variables include whether the facilities are at grade, on elevated structures, or below ground (subways), and consideration of system design principles such as maintaining route continuity, avoidance of excessive convergence of routes, simplification of interchange areas, and avoidance of unnecessary curvature or grades. Therefore, the detailed locational analysis required to accommodate transportation facilities within a dense urban area usually involves unique site-specific considerations such as available vacant land, possible space within existing freeway and other existing transportation facilities rights-of-way, subsurface conditions (in the case of underground routes), and urban design objectives. These detailed considerations are beyond the scope of this book. *Our focus instead is upon estimating the needs in terms of the number of lanes and tracks as a basis for estimating the impacts of a particular alternative for more detailed physical design.*

Capacity and Level of Service (LOS)

In these notes, we adopt the analysis methods of the *Highway Capacity Manual, TRB Special Report 209,* third edition, 1994 (1). Hence, for automobile users, lane density (vehicles per mile, or kilometer) is considered to be the main determinant of LOS. For transit users, LOS is usually more complicated to assess and involves more criteria, such as passenger space within vehicles and extent of loading. Both types of transportation modes are described briefly below. Also, it should be noted that some overlap in the estimated LOS occurs for transit users in the case of bus transit operations on freeways.

Highway Capacity and LOS

The basic relationships between traffic volume and speed of interest for the project described later involve mainline freeway conditions. The relationships between traffic speed, density, volume, and LOS, and the method of calculating the number of lanes required for a given volume of vehicles and LOS, as well as a worked example, are summarized in Figure 3-1. Further details on the definition of LOS are provided in the Appendices.

Basic Traffic Flow Definitions

Speed: Distance traveled per unit time.

Volume: The number of vehicles, trains, passengers etc. passing a given point per unit time.

- Average Daily Traffic (ADT). The total of the volumes in each direction on a highway for a 24-hr period, usually averaged over a period of up to 1-year.
- Design hourly volume (DHV). The total of the volumes in each direction during a selected design hour, usually estimated by a factor, K, or by estimation of the 30th highest hour of a year.

K Factor: The ratio, usually expressed as a percentage, of the DHV/ADT

D: Directional distribution - percentage of traffic traveling in the predominant direction during the design hour.

T: The percentage trucks, or trucks and other heavy vehicles such as buses, in a traffic stream.

Density (or Concentration): The number of vehicles per unit length of highway, usually per lane

Relationships

$DHV = ADT \times K.$ $DDHV = DHV \times D = ADT \times K \times D$

Volume (veh/hr) = Speed (km/hr) x Density (veh/km)

d

Maximum Service Flow Rate Per Lane (Source: based upon Ref. (1)

The relationship linking the traffic flow rates, 'peaking' effects of traffic within a peak hour, adjustments to the traffic flow due to lateral obstructions near the traveled way, lane width, and presence of heavy vehicles and drivers unfamiliar with the location being considered, is given by the equation:

$$MSF_i = \frac{V_i}{PHF \times N \times f_W \times f_{HV} \times f_p}$$ Equation 1

where

MSF_i = maximum service flow rate per hour per lane (pcphpl) for LOS i

N = number of lanes in one direction of the freeway

f_W = factor to adjust for the effects of restricted lane widths and lateral clearances

f_{HV} = factor to adjust for the effect of heavy vehicles on the traffic stream, and

f_p = factor to adjust for the effect of recreational or unfamiliar driver populations

V_i = hourly volume under prevailing conditions (vph), at LOS i

PHF = Peak hour factor

The peak hour factor (PHF) is estimated by the relationship:

PHF = Hourly volume / Peak rate of flow within the hour

Often, the rate of flow within the hour is taken as a 15 minute period, and typical values of the PHF vary between 0.85 and 0.95. The adjustment factors $(f_W, f_{HV}, and f_p)$ are obtained from tabulations based upon observations of driver behavior

Equation 1 may be used as the basis for estimating the operating LOS or the number of lanes appropriate to a given volume. The latter use is the one which will be employed here to estimate the number of lanes of an expressway required to accommodate a given demand as determined from the trip assignment phase described in the previous chapter. The fundamental relationship between vehicle speed, traffic volume and LOS is shown in the diagram on the following page, as well as in tabulated numerical values of the density, volume/capacity ratio, and speeds for each LOS. The table shows the numerical values for a 60 mph [97 km/h] free-flow speed freeway only. An illustration and definitions of Levels of Service are provided in the Appendices.

Figure 3-1   Highway capacity and level of service—Summary of key points

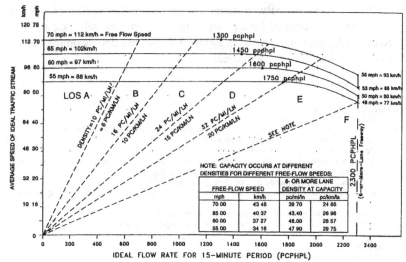

Speed, flow, and level of service criteria (Source: Ref.(1), but including also SI units by author)

LEVEL OF SERVICE (LOS) CRITERIA FOR BASIC FREEWAY SECTIONS, FREE-FLOW SPEED OF 60 MPH [97 KM/H]

LOS	MAXIMUM DENSITY		MINIMUM SPEED		MAXIMUM SERVICE FLOW RATE	MAXIMUM V/C RATIO 4-LANE FREEWAYS	MAXIMUM V/C RATIO 6 OR 8 LANE FREEWAYS
	(PC/MI/LN)	(PC/KM/LN)	(MPH)	(KM/H)	(PCPHPL)		
A	10	6	60	97	700	0.318	0.304
B	16	10	60	97	1,120	0.509	0.487
C	24	15	60	97	1,644	0.747	0.715
D	32	20	57	92	2,015	0.916	0.876
E	41.5/46.0	25.8/46.0	53 0/50.0	85.3/80.5	2,200/2,300	1.000	1.000
F	var	var	var	var	var	var	var

Note:In the table entries with split values, the first value is for 4-lane freeways, and the second is for 6- and 8-lane freeways
Source: Based upon Ref. (1), Table 3-1.

Level of service criteria for basic freeway sections (Source: Ref.(1), but including also SI units by author)

For sketch planning studies of the type illustrated later in this book, the peak hour factor and adjustment factors may be consolidated into a single adjustment factor in order to simplify the calculations.

Example
In order to illustrate the relationships of the foregoing concepts, this simplified numerical example addresses the need to estimate the number of lanes needed for a freeway.

Problem
A freeway having a free-flow speed of 60 mph [97 km/h] is to be designed to operate at LOS C, with an ADT of 60,000 vehicles. The value of K is 10%, and the directional distribution, D, is 75%. The combined value of the PHF and the adjustment factors is estimated to be 0.85. How many lanes wil be needed for the freeway?

Solution
From examination of the table "Level of Service Criteria for Basic Freeway Sections" we see that for LOS C the maximum service flow rate is 1,644 pcphpl. Therefore, inserting the values in Equation 1, we can write:

MSF = 1,644 pcphpl = 4,350 /(0.85 x N)

Hence, N = 3.11 , say 3 lanes, and the total number of lanes (2-way) required for the freeway = 3 x 2 = 6

Figure 3-1 continued

Transit Capacity and LOS

The major elements involved in the estimation of the number of lanes or tracks for public transportation facilities require specification of a number of inputs. A detailed analysis would include consideration of space per person in transit vehicles, travel time, number of vehicles per train, and frequency of service (including walking and waiting time).

Other factors less easily quantifiable in estimating the total transit LOS for a given trip include security within the vehicles, at stops, terminals, park and ride, and at transfer areas. Also, because the use of transit may involve feeder service by transit or automobile, as well as line-haul service, many variables may occur in a user's total trip. In keeping with the main objectives of this book, the focus here will be on the line-haul segment of the trip, because this is usually the most extensive single element of the total trip.

Typical operating values for rail rapid transit (RRT) and for light rail transit (LRT) facilities are shown in Figure 3-2, with an example of estimating the required number of tracks. This also shows the assumed LOS and related transit operating parameters for illustrative purposes, and an estimate of the associated fleet size. The service volumes are seen to vary widely, and the selection of the most appropriate combination of modes depends on the capacity and the nature of the demand and operations. Additional information regarding typical and observed transit vehicle and passenger volumes is provided in the Appendices.

Determining the Mix of Physical Facilities

Planning and design of the system require determination of a mix of modes that accommodates the total demand at the stated levels of service and minimizes costs and adverse impacts.

Method

No generally adopted procedure for estimating the most desirable mix of modes exists. This is largely the result of the mix of modes in a corridor evolving over many years as the transportation system develops in response to city and suburban growth and other socioeconomic influences. However, several authors address the issue to varying degrees (2–8). In order to review and adjust the major variables and to ensure that the computations can be repeated within a reasonable time, a useful series of steps adopted for the examples in this book, and also shown in Figure 3-3, is as follows:

1. Identify the link in the corridor that is carrying the highest passenger volumes in the design hour.

Basic Definitions

Volume and Speed are defined in similar terms to those for automobile traffic, but other highway related terms are generally different for transit, although related. The following self-explanatory relationships may be expressed for transit vehicles operating one-way along a particular route segment:

Passengers/hr = Trains/hr × Cars/Train × Seats/Car × Passengers/Seat, or,
Passengers/hr = Cars/Hr × Seats/Car × Passengers/Seat

Bus Services

Levels of service for bus transit vehicles are as follows (based upon Ref. (1)):

(50-Seat, 340-Sq Ft [211-Sq m] Bus)

Peak Hour Level of Service	Passengers	Approx. max. area/pass Sq ft	Sq m	Approx. area/pass Sq ft	Sq m	Pass/Seat Approx.
A	0 to 26	>13.0	>8,1	>1.5	>0.9	0.00 to 0.50
B	27 to 40	13.0 to 8.5	8.1 to 5.3	1.5 to 1.1	0.5 to 0.7	0.51 to 0.75
C	41 to 53	8.4 to 6.4	5.3 to 4.0	1.0 to 0.8	0.6 to 0.5	0.76 to 1.00
D	54 to 66	6.3 to 5.2	3.9 to 3.2	0.8 to 0.6	0.4 to 0.37	1.00 to 1.25
E (Max. sched. load)	67 to 80	5.1 to 4.3	3.1 to 2.7	0.6 to 0.5	0.37 to 0.30	1.26 to 1.50
F (Crush load)	81 to 85	<4.3	<2.7	<0.5	<0.3	1.51 to 1.60

Source: Based upon Ref. 1, Table 12-5.

For buses operating on arterial and CBD streets, suggested guidelines for different levels of service are as follows (based upon Ref. 1):

SUGGESTED BUS PASSENGER SERVICE VOLUMES FOR PLANNING PURPOSES (HOURLY FLOW RATES BASED ON 50 SEATS)

LEVEL OF SERVICE (STREET)	BUSES/ PASSENGERS/SEAT	LEVEL OF SERVICE (PASSENGERS)				
		A 0.00-0.50	B 0.51-0.75	C 0.76-1.00	D 1.01-1.25	E 1.26-1.50
		ARTERIAL STREETS				
A	25 or less	625	940	1,250	1,560	1,875
B	26 to 45	1,125	1,690	2,250	2,810	3,375
C	46 to 80	2,000	3,000	4,000	5,000	6,000
D	81 to 105	2,625	3,940	5,250	6,560	7,875
E	106 to 135	3,375	5,060	6,750	8,440	10,125
		CBD STREETS				
A	20 or less	500	750	1,000	1,250	1,500
B	21 to 40	1,000	1,500	2,000	2,500	3,000
C	41 to 60	1,500	2,250	3,000	3,750	4,500
D	61 to 80	2,000	3,000	4.000	5,000	6,000
E	81 to 100	2,500	3,750	5,000	6,250	7,500

NOTE: Ratio shown for level of service (passengers) is "passengers per seat" on average bus. Thus 1 00 means 50 passengers for the assumed 50 seats
Values would be 6 percent higher for a 53-seat bus
Values for articulated buses would be 15 to 20 percent greater

Figure 3-2 Transit capacity and level of service—Summary of key points *Source of numerical data:* Ref. (1)

Rail Services

1. Rail Rapid transit (RRT) operating characteristics are shown below, based upon a 2-min headway (30 trains/hr). Current recorded maximums are approximately 50,000 passengers per hour per track, although higher volumes have occasionally been noted. A reasonable LOS for design purposes would be approximately LOS C-D. Assuming 120 passengers/car, 7 car trains, and a headway of 30 sec (30 trains/hr) would result in an hourly passenger volume per track of 26,200 passengers per track, one-way, at an LOS of C-D. This value will be used later herein for design purposes.

Typical Rail Transit Capacities - 30 Trains per Track Per Hour, 2-Min Headway (Flow Rate)

		Car Length		Approx.	Passengers Per Hour				
					0% Standees	50% Standees	100% Standees	150% Standees	200% Standees
Cars/Trains	Cars/Hour	(Ft)	(m)	Seats/Train	(1.00)**	(1.50)**	(2.00)**	(2.50)*	(3.00)**
6	180	50	31	300	9,000	13,500	18,000	22,500	27,000
		75	47	450	13,500	20,250	27,000	33,750	40,500
8	240	50	31	400	12,000	18,000	24,000	30,000	36,000
		75	47	600	18,000	27,000	36,000	45,000	54,000
10	300	50	31	500	15,000	22,500	30,000	37,500	45,000
		75	47	750	22,500	33,750	45,000	56,250	67,500
FT/ Passenger:					10	6.7	5	4	3.3
Passenger Level of Service (U.S & Canada Conditions)					B	C	D	E-1	E-2
Comments:									Maximum schedule loads

* Approximate

Source: Based upon Ref. (1) Table 12-15

2. Light rail transit (LRT) operating characteristics, including also tram or streetcar service, can vary considerably, and are shown below. Assuming 120 passengers/car and 2-car trains operating at a headway of 2 min (30 trains per hour) would result in an hourly passenger volume of 10,800 passengers per hour per track, for LOS D. Note that for on-street operation, even with a reserved right-of-way, a maximum practical volume would be 12 trains per hour to avoid "bunching" and associated disruptions to schedules. Typical ranges in capacities are as follows (Ref. (1)).:

		Pass. Level of Service	
		D	E
	Units Per Hour	5.0 Sq ft [3.1 Sq m] per person	Max. Schedule Loads 3.3 Sq ft [2 Sq m] Per Person
Street Cars: Single 46-50 ft [29-31 m] unit on street.	90	7,500	12,000
LRT-Off street: Three 75 ft [47 m] [47 m] car units.	30	11,000	17,500
	35	13,000	20,000

Current operating experience in the United states and Canada suggest maximum realizable capacities of 12,000 to 15,000 persons per track per hour. However, the European experience shows up to 20,000 persons per hour.

Source: Based upon Ref. (1), Ch. 12.

Figure 3-2 continued

Fleet Size Estimation

Estimation of the number of vehicles required to accommodate the maximum passenger volume along some segment or all of the corridor is necessary in order to estimate the cost of the vehicle fleet.

Normally, the maximum passenger volume in the A.M. peak hour will occur in the inbound direction to the CBD and in the outbound direction (from the CBD) in the P.M. peak hour. Under these conditions, for one direction of travel the vehicles will be underutilized. Nevertheless, the full fleet of vehicles will be required to be present during these times, assuming that the vehicles continue to operate on a continuous "round trip" basis.

The number of times that any given vehicle passes a point on the corridor during a given time period will depend upon its average speed throughout the corridor and the length of the corridor. The average speed includes the time taken for all stops for loading or unloading passengers or waiting for other reasons. For a default value for terminal time, a suggested value is 10% of total running time. If the round trip distance is D_{rt} miles, the average speed for the round trip is V_{av} mph and the average travel time is T_{rt} hr,

$$T_{rt} = D_{rt}/V_{av} = \text{miles}/(\text{miles/hr}) = \text{hr} \qquad [= \text{Km} / (\text{Km/h}) = \text{hr}]$$

And the number of evenly spaced vehicles, N, passing any point in the corridor in a given time T at a rate of q veh/hr may be stated as
$$N = qT = \text{veh/hr} \times \text{hr} = \text{veh}.$$

and $q = N/T$

Thus the number of vehicles required in the fleet to pass the given point N times in time T is

$F = N \times$ the number of times the vehicle makes the round trip per hour
$$= N(T_{rt}/T) = T_{rt}q$$

Example

The peak load point on an inbound RRT corridor is 30,000 passengers per hour. Each train of 6 cars operating on the route can carry 2,000 passengers. If the corridor length is 6 miles [9.7 km] and the average train speed is 20 mph [32 kph], how many trains are required in the fleet and what will be the headway in minutes?

Time for the round trip, $T_{rt} = D_{rt}/V_{av} = (2 \times 6 \text{ ml})/20 \text{ mph}$ $= (2 \times 9.7 \text{ km}) /32 \text{ km/h} = .067$ hr

Train volume required to accommodate 30,000 pass/hr $\quad = q = 30,000$ pass/hr $/ 2,000$ pass/train
$= \underline{15 \text{ trains/hr}}$

Fleet size, $F = qT_{rt} = 15 \times 0.67 = 10$ trains $= 10 \times 6$ cars/train $= \underline{60 \text{ RRT cars}}$

Headway $= 1$ hr $\times 60$ minutes/hr $/ 15$ trains/hr $= \underline{4 \text{ minutes}}$

Figure 3-2   continued

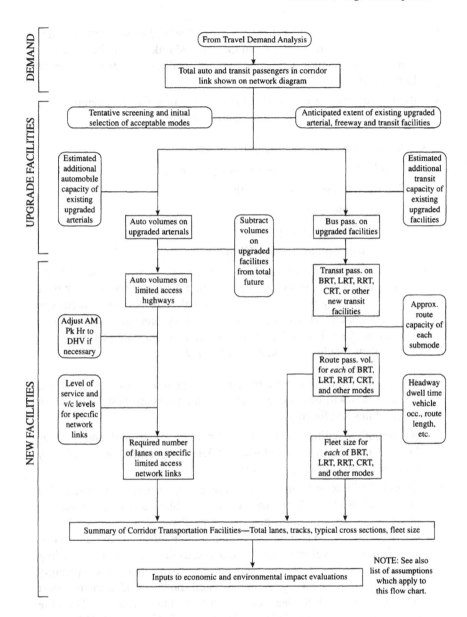

Figure 3-3   Facilities estimation—Flow diagram

2. Identify and define the available service flow and capacity of the existing upgraded facilities, if any, on the selected link, so that the remaining demand may be allocated to the planned facilities. See the notes about LOS in the preceding section.

3. Broadly define the candidate modes in the mix for consideration. For most large cities these would normally include:

   - Automobiles on arterials and freeways

   - Regular buses on arterials and freeways (RB)

   - Bus rapid transit, or busway (BRT)

   - Light rail transit (LRT)

   - Rail rapid transit (RRT)

   - Regional or commuter rail (RR or CR)

   - Ferry, bicycle, or other modes, depending on their availability. However, modes transporting less than, say, 5% of the total corridor peak period one-way passenger volumes usually are not major factors in affecting the levels of impacts or total capital cost at the preliminary analysis stage. If significant nonmotorized modes are possible, however, they should be included at this stage.

4. Modify the future adjusted directional design hourly volumes (DDHVs) to those that will be accommodated by the proposed new facilities for automobiles and for transit.

5. Automobile facilities—estimate the number of arterial and freeway lanes at the design LOS (adjusted DDHV for automobiles divided by service volume at the design LOS).

6. Transit facilities—various scenarios involving one or a combination of BRT, LRT, RRT, CRT, or other kinds of public transportation may be appropriate. See the Appendices for typical ranges of volumes by the different modes. Maximum and minimum frequencies and associated vehicle and train volumes must be defined, as described in the preceding section. CRT would not be used unless existing track and operating facilities are available. Selection of each mode would normally start with the mode with the least capital cost and lowest capacity. Thus, the order would usually be (1) BRT, (2) LRT, (3) RRT, (4) CRT. Other general guidelines (i.e., before the costs and impacts described in the next section have been reached) attempt to attain full utilization of each mode in terms of matching the specified maximum vehicle volumes with the specified line capacity, or service flow. Another consideration is whether a rail system now exists or is being extended, possibly to serve an outlying major generator such as an airport. The final decision

regarding a choice of mode may depend upon specific characteristics of the area and the fact that it is usually advisable to have both bus and rail service of some sort, in order to build some redundancy into the system. Thus, if an areawide electrical power failure or serious accident occurs on the rail facilities, passengers could then be transported by buses. The final choice of system will depend upon the detailed analysis of modal capacities and operating characteristics. However, the order of priority noted above provides a basis for developing further alternatives in greater detail.

Using the above guidelines, Figure 3-4 provides an example of estimating a possible mixture of modes that would follow the determination of passenger volumes for automobile and transit modes for a link in the corridor routes during the traffic assignment phase of the demand analysis. This example is shown for peak hour movements in the predominant direction on the link with the maximum volume.

Selection of the Preferred Alternative

An important question that arises at this stage is, Which of the alternatives is preferred? To answer this question, the impacts of each must be examined in detail, and a wide variety of economic, financial, and environmental factors must be considered, not only within the corridor itself but also in consideration of the effects on the regional transportation. At this stage, however, the definition of alternatives can only be guided by the general principles noted above, and the matter of selecting the preferred alternative is presented under the description of cost estimates in the next section.

Outputs

The outputs of this phase are primarily the number and extent of fixed facilities (number of highway lanes, tracks of rail transit by category), operating characteristics of each mode such as travel speed, frequency, or headway, and the fleet sizes for each of the modes. These outputs are the basic requirements for estimating the impacts described in the next phase.

Impact Estimation

Estimates of the impacts of a proposed project provide the quantitative results associated with each of the specified policies, ranging from automobile intensive to transit intensive. In addition to an economic analysis, the impact estimates provide the basic information for deciding if a proposed project meets available

Problem Statement

It is required to design the line-haul transportation facilities (number of lanes and tracks inbound) to accommodate 32,000 transit passengers and 12,600 automobiles per hour inbound to the CBD in the A.M. peak hour. The public transportation passengers are to be accommodated on a mixture of bus and rail facilities in order to provide some "redundancy" to the overall system should one or the other mode experience a serious failure. Provide three alternative plans which could be used for more detailed analysis and which offer different levels of service for the automobile users - a preferred LOS C and a capacity LOS E. The estimates may be used later to estimate fleet sizes, costs, air pollution, and energy consumed in the A.M. peak hour.

Input values

At this stage typical "default" values for the design parameters are to be used; these values are as follows:

ITEM	MODE		
	Bus Rapid Transit (BRT)[1]	Light Rail Transit (LRT)	Rail Rapid Transit (RRT
Transit			
Volume, veh/hr	180	90	90
Veh/train	1	3	8
Headway, sec	20	120	120
Volume, trains/hr	NA	30	30
Veh occ, pass/veh	50	150	120
Veh occ, pass/train	NA	450	960
Volume, pass/hr	9,000	13,500	28,800
Travel speed, mph	40	20	25
Travel speed, kph	65	32	40
Approx LOS	B-C	D	D
Other conditions	Max 1 lane	Max 1 track	Max 1 track
Automobile	LOS C	LOS E	
Volume, veh/hr	1,200	2,200	
Veh occupancy, pass/veh	1.2	1.2	
Volume, pass/hr	1,440	2,640	

(1) The volume of buses may depend on the extent of terminal facilities in the central city area.

Solution

The method adopted here for estimating a combination of modes that accommodates the required demand is illustrated on the following page for a hypothetical corridor.

For each of the three alternatives, the approach adopts progressive steps from the initial statement of demand, the estimation of the number of highway lanes needed based upon the highway LOS criteria, and then the allocation of transit demand to BRT, LRT, and RRT. For the transit facilities, the three alternatives are as follows:

Alternative 1 BRT - The maximum amount of passengers (12,000) are allocated to this mode. LRT -The 20,000 remaining transit passengers could not all be accommodated on this mode, and a decision is made to offer no LRT services.   RRT - The remaining 20,000 passengers are allocated to this mode, resulting in a v/c ratio of 0.79 with one track, with no allocation to LRT. This alternative would be appropriate where BRT services are consistent with the regional highway

Figure 3-4   Combining transportation modes—Example

system, where a relatively narrow corridor of high demand exists to ensure patronage of the RRT, or where feeder bus services channel high volumes of passengers to the RRT system.

Alternative 2   This alternative provides a more diverse transit system in that the BRT, LRT and RRT are used to accommodate the total transit passengers. This results in lower v/c ratios than Alternative 1, but would provide for a more scattered source of passengers, and, as is usual with LRT, provide more stops along the route than would RRT.   However, three modes and the lower v/c ratios may result in higher capital costs and costs per passenger mile.

Alternative 3 is the same as Alternative 2, but automobile traffic operates at LOS E (capacity), instead of LOS C (with Alternative 2), and the split between LRV and RRT is slightly different.

Comments
The results show how three possible alternatives can accommodate the passenger demand within the overall constraints of headway, vehicle occupancy and frequency. Clearly, other combinations of modes are possible. It is also important to note that in the example, the modal split has been held constant. However, because the change in LOS from C to E implies a greater travel time for automobile users it is likely that the volume of automobile users would decrease somewhat and the numbers of transit riders increase. This change may be calculated in an iterative fashion until an equilibrium is reached. Other modifying factors also occur in practice, and the decision about how far to refine the estimates ultimately rests on judgment and a realization that operating conditions may vary for a variety of reasons beyond the ability of the analyst to estimate or the operator to influence.

COMBINING MODES - THREE ILLUSTRATIVE ALTERNATIVES

MODE	ITEM	UNITS	ALTERNATIVE 1 VALUE	COMMENTS	ALTERNATIVE 2 VALUE	COMMENTS	ALTERNATIVE 3 VALUE	COMMENTS
Design Year Demand								
	Trans. pass.	Pass/hr	32,000 00		32,000.00		32,000.00	
	Auto vol	pcph	12,600 00		12,500.00		12,600 00	
Autos								
	Volume	pcph	12,600.00	Total auto pass	12,600 00	Total auto pass	12,600.00	Total auto pass
	Service vol	pcphpl	1,200 00	LOS C	1,200.00	LOS C	2,000.00	LOS E
	La. req'd	Number	10 50	Say 10 la	10 50	Say 10 la	6 30	Say 6 la
Bus RT (SRB)								
	Volume (pass.)	Pass/hr	9,000.00		6,000.00		6,000.00	
	Veh occ	Pass/veh	50 00		50.00		50.00	
	Volume(veh.)	Veh/hr	180 00		120.00		120.00	
	Service vol	Veh/hr	180 00		180.00		180 00	
	La. req'd	Number	1 00	Say 1 la	0.67	Say 1 la	0 67	Say 1 la
Transit pass remaining			23,000.00		26,000.00		26,000.00	
LRT								
	Volume (pass.)	Pass/hr	0.00		9,000 00		8,000.00	
	Train occ	Pass/train	300.00		300.00		300.00	
	Volume(trains)	Train/hr	0.00		30 00		26 67	
	Service vol	Train/hr	30 00		30.00		30.00	
	La. req'd	Number	0.00		1.00	Say 1 track	0 89	Say 1 track
Transit pass remaining			23,000 00		17,000 00		18,000.00	
RRT								
	Volume (pass.)	Pass/hr	23,000 00		17,000.00		18,000.00	
	Train occ	Pass/train	840.00		840.00		840 00	
	Volume(veh.)	Veh/hr	27.38		20.24		21.43	
	Service vol.	Veh/hr	30.00		30.00		30 00	
	La. req'd	Number	0 91	Say I track	0.67	Say I track	0 71	Say I track
Transit pass remaining			0		0		0	

Figure 3-4   continued

funding, air pollution, and energy use mandates. If the project is acceptable, the estimates of the impacts assist in incorporating it in funding programs leading to prioritization and implementation within the context of wider urban policy goals.

Estimates of user benefits and other impacts would typically include a wide range of analyses in accordance with appropriate environmental impact reports or other formal statements. Also, the outcomes of the analyses for the alternative plans often provide the information necessary for modifying the proposals before the final design is conducted. The areas for which impacts are often estimated include capital cost of the system, operational cost, user benefits and costs, air pollution emissions, energy use, and other impacts on the natural and man-made environments. In these notes, because the emphasis on illustrating how basic impacts may be estimated and used to compare the alternatives, and because some impacts such as noise pollution are highly affected by the physical location of the facility related to receptors, only a selection of the major factors are examined. These factors are:

- Capital, or construction cost (including purchase of vehicles) of facilities
- Energy requirements for peak hour operations
- Air pollution (carbon monoxide for illustrative purposes) for peak hour operations.

Key aspects of the above impacts and a method of assembling the relevant information are discussed in the following sections.

Approaches to Investigating Impacts

If the objective were to minimize one of the impacts (cost per passenger mile [km], air pollution, or energy use) for a particular alternative, in order to examine the effects on the other two impacts, a total of three different options would be required, as follows:

Approach A
    Minimize capital cost per vehicle mile [km]
        Resulting air pollution impacts
        Resulting energy use impacts
Approach B
    Minimize air pollution impacts
        Resulting capital cost per vehicle mile [km]
        Resulting energy use impacts
Approach C
    Minimize energy use impacts
        Resulting air pollution impacts
        Resulting capital cost per vehicle mile [km]

Clearly, as well as minimizing any one of the impacts, threshold levels could be set for any combination of them to determine which option could best satisfy the requirements of the area being considered. The problem may be amenable to solution and further investigation with mathematical programming methods, although practical considerations and the presence of variables that are difficult to quantify often complicate such approaches to a point where the results have limited practical value.

The steps adopted for illustrative purposes in Chapter 4 reflect Approach A, above, and are as follows:

1. Develop at least two mode combinations (this was done earlier, in the section entitled Determining the Mix of Physical Facilities).
2. Estimate which mode combination has the least capital cost per passenger mile.
3. Proceed with the estimation of the air pollution and energy use impacts of the mode combination selected in Step 2.

If either of the other two options were to be investigated (i.e., minimize air pollution or energy use), the combination of modes would be adjusted for each of the impacts in accordance with the notes on each impact discussed earlier in this chapter.

Facilities Cost

Capital Costs of Fixed Facilities and Vehicle Fleets

Background

The total cost of transportation service includes capital cost, operation and maintenance costs, and user costs. For most economic analyses, all costs are reduced to an annual basis and comparisons made in terms of net present worth, cost benefit, internal rate of return, or some other basis.

In the project illustrated in this book, the initial capital cost is used as the criterion for screening of alternatives, because the objective is to conduct a preliminary screening only. Capital costs are important in the initial screening phase because they provide an indication of the funding required for inclusion in programmed capital improvements, including projects in other infrastructure sectors besides transportation. Thus, the initial financial feasibility (and not the full economic feasibility) is being used as the criterion for initial selection for further, more detailed analysis.

Not including operating costs in the initial analysis tends to result in bus services appearing more attractive because the relatively high labor cost per passenger mile of bus operation is ignored. However, because of the many variables encountered in practice, including consideration of vehicle passenger load factors throughout the operating day and the fact that at least some combination of bus and rail service is desirable for a balanced system, the capital cost only criterion is used for illustrative purposes in Chapter 4.

Data

The unit costs of the various facility elements used in this book are based upon *Characteristics of Urban Transportation Systems (CUTS)*, Urban Mass Transportation Administration, Washington, D.C., 1979, 1985, and 1992 (9–11). Other useful data sources are publications by the Institute of Transportation Engineers (12, 13). They include the construction costs of the way (e.g., freeways), and rapid transit track and control equipment, as well the cost of the public transportation vehicles such as buses and rail cars. Each of the capital cost items and their unit costs, together with an example of their applicability, are shown in Table 3-1.

Method

Because of the preliminary nature of this analysis, estimation of the capital costs will be performed by a simple multiplication of the project's quantities by the associated unit costs. The quantity of each item is measured from the project alternative being considered, as follows:

- Freeway lane miles [km] = length of route × number of lanes.
- BRT lane and LRT, RRT, and CR route miles [km] are measured similarly. A route is defined here as two tracks, one inbound to, and the other outbound from, an assumed central city area.
- BRT, LRT, RRT, and CR vehicle fleet sizes (number of vehicles to accommodate all the passengers on the route in the peak hour) must be estimated based upon the maximum passenger volume on the route, travel speed, and route length. See the earlier description of the method and an example of estimating fleet size in Determining the Mix of Physical Facilities.

The estimated capital cost for each item is calculated and added to provide a total capital cost for the project. Subtotals for each mode and for highways and public transportation are usually listed in a summary format to permit a clear comparison between the cost components for various alternatives. An example of estimating capital costs is shown in Table 3-2. Also shown is the capital cost per passenger kilometer. Although we will still occasionally use both English and SI units, at this point we will predominantly use SI units in our calculations.

Table 3-1　Capital Costs for Transportation Facilities, and Example

Facility Type	Item	Units	Value	Comments
Freeways	Unit Cost	$/lane mi	2,200,000	Ref. (11), Table 4-16, illustrative,
		$/lane km	1,366,200	updated to 1995.
BRT	**Way:**			
	Unit cost	$/lane mi	2,200,000	Ref. (10), Table 3-14, illustrative,
	Unit cost	$/lane km	1,366,200	updated to 1995.
	Fleet:			
	Unit Cost	$/bus	190,000	Ref. (11), Table 3-11.
	Stations:			
	Unit Cost	$/station	1,000,000	Ref. (11), Table 2-17, illustrative, similar to LRT station.
LRT	**Way:**			
	Unit Cost	$/route mi	20,000,000	Ref. (11), Table 2-15, illustrative,
		$/route km	12,420,000	at grade (includes stations).
	Fleet:			
	Unit Cost	$/LRT vehicle	1,600,000	Ref. (11), Table 2-20, illustrative.
	Stations:			
	Unit Cost	$/station	1,000,000	Ref. (11), Table 2-17, illustrative.
RRT	**Way:**			
	Unit Cost	$/route mi	70,000,000	Ref. (10), Table 2-25, illustrative,
		$/route km	43,470,000	at grade (includes stations).
	Fleet:			
	Unit Cost	$/RRT vehicle	1,500,000	Ref. (11), Table 2-19, illustrative.
	Stations:			
	Unit Cost	$/station	4,500,000	Ref. (9), Table 2-20, illustrative, at grade, updated to 1995.

English to SI conversion: mi to km 0.621

Example

An 8 mile [12.9 km] LRV line (2-tracks, one inbound and one outbound) is to be constructed, including stations. An estimated fleet size of 42 vehicles (cars) will be required. The track is to be constructed at grade for the entire length. Calculate the total capital cost.

Solution

English units:

(Track: 8 mi × $20,000,000/mi) + (Vehicles: 42 cars × $1.5 mill/car) = $160 mill + $63 mill = $223 million

SI units:

(Track: 12.88 km × $12,420,000/km) + (Vehicles: 42 cars × $1.5 mill/car) = $160 mill + $63 mill = $223 million

(Results may differ slightly due to rounding)

Table 3-2　Estimated Capital Cost and Cost per Passenger Kilometer

| | | PROJECT CAPITAL COST | | COST PER PASS. KM | |
ITEM	UNITS	CAPITAL COST, $ (1)	COMMENTS	PASSENGERS (2)	CAP. COST/ PASS. KM.
FREEWAYS					
Length,1-way	km	8	Project length		
Lanes,1-way	number	8	From Lane / Track Requirements		
Total lane length	lane km	128	Length x Lanes x 2-way		
Unit cost	$ / lane km	1,366,200	Table 3-1		
TOTAL COST, FWs	$	174,873,600	Lane Miles x Unit Cost	12,000	14,573
BUS RAPID TRANSIT (BRT)					
Way:					
Length	km	8	Project length		
Lanes	number	1	From Lane / Track Requirements		
Total lane length	lane km	16	Length x Lanes x 2-way		
Unit cost	$ / lane km	1,366,200	Table 3-1		
Cost	$	21,859,200	Lane km x Unit cost		
Fleet:					
Frequency	bus / hr	120	From Lane / Track Requirements		
Average speed	km/h	40	Assumed		
Distance	km	16	2 x Length (round trip)		
Cycle time	hours	0.40	Distance / Average Speed		
Fleet	number	48	Frequency x Cycle Time		
Unit cost	$ / bus	190,000	Table 3-1		
Cost	$	9,120,000	Fleet x Unit cost		
Stations:					
Stations	number	3	Assumed		
Unit cost	$ / station	1,000,000	Table 3-1		
Cost	$	3,000,000	Stations x Unit cost		
TOTAL COST, BRT	$	33,979,200		6,000	5,663
LIGHT RAIL TRANSIT (LRT)					
Way:					
Length	km	8	Project length		
Routes	number	1	From Lane / Track Requirements		
Distance	Route km	8	Length x lines		
Unit cost	$ / line km	12,420,000	Table 3-1		
Cost	$	99,360,000	Track km x Unit cost		
Fleet:					
Frequency	trains / hr	34	From Lane / Track Requirements		
Average speed	km/h	30	Assumed		
Distance	km	16	2 x Length (round trip)		
Cycle time	hours	0.53	Distance / Average Speed		
Cars/train	number	2			
Fleet	number	36	Frequency x Cycle Time x Cars/train		
Unit cost	$ / train	1,600,000	Table 3-1		
Cost	$	58,026,667	Fleet x Unit Cost		
Stations:					
Included in way cost					
TOTAL, LRT	$	157,386,667		10,000	15,739
RAIL RAPID TRANSIT (RRT)					
Way:					
Length	km	0	Project length		
Routes	number	1	From Lane / Track Requirements		
Distance	Route km	0	Length x Tracks		
Unit cost	$ / line-km	43,470,000	Table 3-1		
Cost	$	0	Track Miles x Unit Cost		
Fleet:					
Frequency	trains / hr	0	From Lane / Track Requirements		
Average speed	kph	30	Assumed		
Distance	km	0	2 x Length (round trip)		
Cycle time	hours	0.00	Distance / Average Speed		
Cars/train	Number	0			
Fleet	number	0	Frequency x Cycle Time x Cars/train		
Unit cost	$ / train	1,500,000	Table 3-1, illustrative		
Cost	$	0	Fleet x Unit Cost		
Stations:					
Included in way costs.					
TOTAL, RRT	$	0		0	NA
TOTAL PROJECT	$	366,239,467		28,000	13,080

Notes:
(1) All input values, shown in boxes, are assumed, for illustrative purposes
(2) Numbers of passengers, shown in boxes, are assumed, for illustrative purposes.

Capital Cost per Passenger Mile [Km]

Basic objective

The objective is to provide some indicator of the relative cost of each of the modal mixes proposed for each of the alternatives. One way of doing this is to compare the number of people served per unit of time and cost. This is one measure of the cost effectiveness of the system. Normally, as indicated earlier, a cost indicator of this type would include consideration of capital costs, maintenance, operating costs, and user costs, all reduced to a "per passenger mile [km]" basis. However, for the purposes of these notes an approximate surrogate for comparing *between* alternatives, rather than an *absolute* estimate of economic efficiency, is the estimate of the capital cost per passenger mile [km] per hour. The selected period during which these distances are traveled is the peak hour, because this period is close to the design hour in most instances. Therefore, the indicator may be expressed as:

capital cost per passenger mile [km] = capital cost/(total peak hour passengers
$\times$ corridor length $\times$ 2).

This expression represents the capital cost of the route, i.e., two-way fixed and rolling equipment, divided by two to represent the one-way capital cost, all divided by the inbound passenger volume. This method maintains the same number of inbound passengers to simplify the calculations.

For any given passenger distance, the capital cost will be affected as follows:

1. The more passengers that can be accommodated on the lower capital cost transit systems (i.e., BRT and LRT), the lower will be the total capital cost per passenger kilometer.
2. The greater the lane or track utilization of a specific mode, the lower will be the capital cost per passenger kilometer. For example, rounding up from 3.6 to 4.0 lanes of freeway will result in a greater cost per passenger mile [km] than rounding up from 3.9 to 4.0. Also, a 0.67 utilization of light rail vehicle (LRV) facilities will result in a greater cost per passenger mile [km] than 0.85.

An example of the estimated capital costs per passenger kilometer is shown in the last column of Table 3-2, preceded by the estimate of capital costs for each item. Further examples of capital costs, and capital costs per passenger kilometer, are discussed in Chapter 4.

Energy Use

Background

Beginning in the early 1970s and later in the early 1980s the United States began to include energy consumption in the assessment of environmental impacts. Although a comparative total surplus of oil exists worldwide (1993–94) with pockets of shortages due to economic or political pressures, questions of global resource sustainability and the very real adverse impact on the immediate environment mandate the need for continued conservation of energy.

Transportation Energy Use and Measurement

Total energy requirements for a transportation project consist of several sources of fossil fuels or, as is sometimes the case, hydro or nuclear power. For highway modes, including buses operating on freeways, gasoline or diesel fuel provides the basic source of energy. For electrically powered rail and trackless trolley (trolley bus) vehicles, the basic source may be residual oil, coal, nuclear, or hydro power used in a generator and transferred to an electrical grid system. Because of these varied sources it is necessary to convert each to a common measure in order to compare, for example, liters of gasoline per automobile kilometer with liters of residual oil per RRT car kilometer. Conversion factors between the several measures of energy are provided in the Appendices. The joule (J), kilo, and megajoule (kJ, MJ) are commonly adopted measures. In general, MJ provides a convenient, common unit for energy use in the examples that follow.

Data Needs

Numerous sources of fuel consumption data exist, and these are updated as vehicle and fuel technology are improved. The fuel efficiency of most private vehicles is increasing. However, the average fuel consumption per vehicle in the nation has recently (late 1980s–1990s) increased because of the preference of consumers for light trucks, utility vehicles, and larger automobiles.

For automobiles, the relationship of energy consumption to vehicle speed and ambient temperatures is obtained from empirically derived data often portrayed in graphs, such as Figure 3-5. Fuel consumption of buses and rail vehicles may be similarly estimated. For rail vehicles, where electrical power is derived mostly from oil- or coal-powered generating stations, fuel consumption may also be estimated. A listing of unit fuel consumption values for the various modes is also provided in the Appendices.

Method

For most purposes where a comparative, approximate estimate is being made, one method is to employ typical unit values of fuel consumption for the vehicles

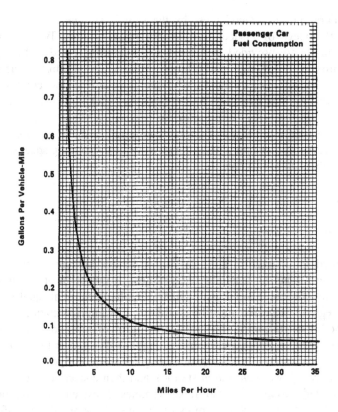

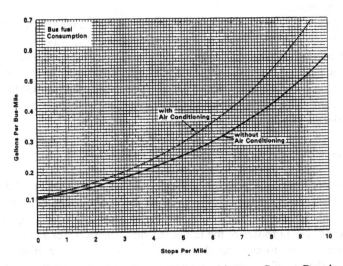

Figure 3-5 Energy consumption of automobiles and buses   *Source*: Based on data in Ref. (14)

under consideration, and extend these by the vehicle volume at a given speed (linked directly to level of service) over the appropriate distance. Typical data and a worked example are provided in Table 3-3. It should be noted that at speeds between approximately 25 mph and 60 mph [40km/h and 97 km/h]—the speed range typically experienced by most traffic operating in free-flow conditions on urban freeways—the fuel consumption does not vary greatly. This is reflected in the data presented for this example. For cases where a wider range of speeds exists, more detailed unit values of energy consumption should be used. These are listed in the Appendices for a wider range of vehicle operating speeds, typical of those described later in Chapter 6.

Air Pollution

Background

As described in Chapter 1, air pollution emissions are one of the critical aspects of determining the feasibility of a project. If operation of the proposed project does not contribute to the maintenance or attainment of air quality specified for the region, the project must be modified or its implementation reconsidered. All modes of transportation emit some form of pollution, whether the source is fuel used in internal combustion engines directly or through diesel, residual oil, or coal used in turbines or other engines at a power generation station (more accurately a power *conversion* station), and then converted into electrical energy distributed through a grid to transit vehicles. These emissions, modified by the climatic and topographic features of the area, have an adverse effect on sensitive receptors.

Regardless of the source, a comparison of the emissions for a particular system will require data on the passenger and vehicle modes of transportation, ambient air temperatures, air pollution controls on the vehicle, travel speed, number of stops and starts, and numerous other factors depending upon the amount of time the engine has been operating since start up. In the projects described here and in Chapter 4, unit values of air pollution are used. These would have to be modified as the project design is refined and proceeds to completion.

Transportation Air Pollution Measurement

Estimation of the impacts of air pollution involves consideration of the nature and extent of the emissions, and the climatic, topographic, and other features of the region that affect their transportation and dispersal before reaching the receptor of interest. For purposes of the initial feasibility analysis for a given region, it is usually sufficient to consider the nature and extent of the emissions, on the assumption that the dispersion will be approximately the same for most projects built in the same location.

Table 3-3   Energy Consumption of Urban Transportation Modes, and Example

Item	Units	Value	Sources and unit conversions
Autos on limited access highways			
Gasoline	gal/veh-mi	0.05	Ref. (11), Table 4-7, includes some trucks
	l/veh-km	0.13	2.53 l/veh-km = 1 gal/veh-mi
	Btu/veh-mi	6,250.00	125,000.00 Btu/gal.
	MJ/km	4	0.0006557 MJ/km = 1 Btu/mi
Buses on freeways (no priority)			
Diesel	gal/veh-mi	0.28	Ref. (11), Table 3-20.
	l/veh-km	0.71	2.53 l/veh-km = 1 gal/veh-mi
	Btu/veh-mi	39,200.00	140,000.00 Btu/gal.
	MJ/km	26	0.0006557 MJ/km = 1 Btu/mi
BRT			
Diesel	gal/veh-mi	0.28	Ref. (11), Table 3-20.
	l/veh-km	0.71	2.53 l/veh-km = 1 gal/veh-mi
	Btu/veh-mi	39,200.00	140,000.00 Btu/gal.
	MJ/km	26	0.0006557 MJ/km = 1 Btu/mi
Rail facilities on exclusive right-of-way			
LRT			
Residual oil	gal/veh-mi	0.35	Ref. (10), Table 2-15, residual oil, updated value.
	l/veh-km	0.89	2.53 l/veh-km = 1 gal/veh-mi
	Btu/veh-ml	49,000.00	140,000.00 Btu/gal.
	MJ/km	32	0.0006557 MJ/km = 1 Btu/mi
RRT			
Residual oil	gal/veh-mi	0.44	Ref. (10), Table 2-13, residual oil, updated value.
	l/veh-km	1.11	2.53 l/veh-km = 1 gal/veh-mi
	Btu/veh-mi	61,600.00	140,000.00 Btu/gal.
	MJ/km	40	0.0006557 MJ/km = 1 Btu/mi

Example

A peak hour, two-way volume of 3,400 automobiles (including a small percentage of trucks), 200 diesel buses (no priority), and 20 RRT trains of 7 cars each is traveling along a 6 mile [9.66 km] segment of an urban corridor. The highway vehicles are operating at LOS C. Estimate the total energy used by the vehicles in this corridor during the peak hour.

Solution

		Distance		Unit energy use		Total energy use	
Mode	Veh. volume	mi	km	Btu/ml	MJ/km	Btu	MJ
Autos and trucks	3400	6	9.66	6,250.00	4	127,500,000	134,599
Buses	200	6	9.66	39,200.00	26	47,040,000	49,659
RRT	140	6	9.66	61,600.00	40	51,744,000	54,625
Total						226,284,000	238,883

Data Needs

As with energy use, the relationship of pollution emissions to vehicle operation and ambient temperatures is obtained from empirically derived data. Selected relationships of automobile air pollution emissions to vehicle operating variables such as running speed, ambient air temperature, and other factors are shown in Figure 3-6. In general, as the speed of a vehicle increases from a crawl to approximately 20 mph [32 km/h], carbon monoxide and hydrocarbon emissions decrease considerably, and nitrogen oxide emission increases.

As when estimating energy requirements, the source of the fuel must be taken into account. Therefore, emissions data for power stations related to distance traveled by the various forms of rail transit or trackless trolleys must be obtained. In the conversions described in this book, it is assumed that electricity generating stations use residual oil as the basic power source.

Method

For most purposes where a comparative, approximate estimate of air pollution emissions typical of sketch planning approaches is being made, one method is to employ unit values of the emissions for the vehicles under consideration and extend these by the vehicle volumes at a given speed (linked directly to level of service) over the appropriate distance. Data and a worked example illustrating the application of average air pollution factors for carbon monoxide (CO) emissions are provided in Table 3-4.

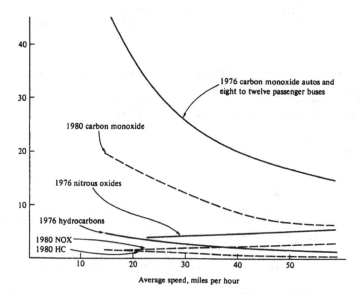

Figure 3-6 Air pollution emission characteristics   *Source*: Ref. (15)

Table 3-4 CO Air Pollution Emissions of Urban Transportation Modes, and Example

Item	Units	Value	Comments (1)
Autos on limited access highways			
CO Emissions Factor	gm/veh mi	6.55	Ref. (11), Table 4-17, LOS E
	gm/veh km	4.07	
CO Emissions Factor	gm/veh mi	14.49	Ref. (11), Table 4-17, LOS C
	gm/veh km	9.00	
Buses on freeways (no priority)			
CO Emissions Factor	gm/veh mi	13.44	Ref. (11), Table 4-17, LOS E
	gm/veh km	8.35	
CO Emissions Factor	gm/veh mi	16.41	Ref. (11), Table 4-17, LOS C
	gm/veh km	10.19	
BRT			
CO Emissions Factor	gm/veh mi	16.41	Ref. (11), Table 4-7, LOS C
	gm/veh km	10.19	
LRT			
CO Emissions Factor	gm/veh mi	0.0100	Ref. (10), Table 2-17, residual oil,
	gm/veh km	0.0062	rounded
RRT			
CO Emissions Factor	gm/veh mi	0.0100	Ref. (10), Table 2-16, residual oil,
	gm/veh km	0.0062	rounded

(1) Notes:

1 km = 0.621 mi

LOS C for a 60 mph [97 km/h] free-flow freeway occurs at 60 mph [97 km/h] (Ref. (1)).

LOS E for a 60 mph [97 km/h] free-flow freeway occurs at 50 mph [80 km/h] (Ref. (1)).

Example
The same vehicle mix and volumes as for the previous example of estimating energy use, i.e., a peak hour, two-way volume of 3,400 pcphpl (including a small percentage of trucks), 200 diesel buses, and 20 RRT trains of 7 cars each is traveling along a 6 mile [9.66 km] segment of an urban corridor. The highway vehicles are operating at LOS C. Estimate the CO emitted by the vehicles in this corridor during the peak hour.

Solution

Mode	Veh. Volume	Distance mi	Distance km	Unit Emissions gm/veh mi	Unit Emissions gm/veh km	Total Emissions (Vol × Dist × unit emiss.) gm
Autos and trucks	3,400	6	9.66	14.49	9	295,596
Buses	200	6	9.66	16.41	10.19	19,692
Rail rapid transit (RRT)	140	6	9.66	0.01	0.01	8
Total						315,296

Review

The nature of the transportation system and the resulting impacts have been shown to be shaped by a combination of the demand for transportation, the technological and performance characteristics of the various modes used in the planning and design of the system, and the unit values of costs, energy use, and air pollution associated with operation of the vehicles.

Starting with the concept of level of service (LOS), i.e., a measure of the quality that a particular service offers to users, the associated lane volumes of automobiles, frequency of service and passenger space on transit vehicles, the abilities of the individual modes to accommodate the volumes of passengers, and the method of estimating the required lanes of highway or tracks of rail transit were determined. A procedure for calculating the combination of modes required, as might be the case of the design for a key link in a transportation corridor, was then presented. This reinforced the notion that several combinations of modes could meet the demands in a particular corridor, and that considerations of modal redundancy, previous experience of the operating agency with particular modes, or other preferences could influence the decision about which mode combinations should be considered as candidates for more detailed investigation.

Examples of estimating the impacts resulting from the various modes were presented. These estimates were based upon simple unit values of the capital costs, energy use, and air pollution, multiplied as appropriate by the highway lane or transit track length, fleet size, number of stations, and related features of the design alternative being considered. It is recognized that these unit values provide only approximate results and do not take into account changes in vehicle speeds, operation for different periods of time, user and external community costs, and other factors that should be considered in a more detailed design and impact estimation procedure.

Service policy variables were incorporated into the supply analysis primarily by means of specifying the LOS for highway users. No explicit LOS was stated for transit users. However, selection of the passenger volumes at well below the bus lane or transit track capacities in the examples presented would ensure that an adequate transit LOS was available (minimal numbers of standees), and such features as the number of vehicles and frequency of service could be readily adjusted and refined when required.

The approach to demand forecasting in Chapter 2, together with the design and impacts estimation methods of this chapter, provide basic guidelines for conducting a practical corridor impact study—the subject of Chapter 4.

References

1. Transportation Research Board, National Academy of Sciences, *Highway Capacity Manual,* Special Report 209, third edition, Washington, D.C., 1992.

2. Armstrong-Wright, A., *Urban Transit Systems, Guidelines for Examining Options,* World Bank Technical Paper Number 52, Washington D.C., 1986.

3. Gray, G.E., and L.A. Hoel, eds., *Public Transportation,* Prentice-Hall, Englewood Cliffs, NJ, 1992.

4. Levinson, Herbert S., Crosby L. Adams, and William F. Hoey, *Bus Use of Highways: Planning and Design Guidelines,* NCHRP 155, Transportation Research Board, Washington, D.C., 1975.

5. Boris S. Pushkarev and Jeffrey M. Zupan, *Public Transportation and Land Use Policy,* A Regional Plan Association Book, Indiana University Press, Bloomington, IN, 1977.

6. U.S. Department of Transportation, Urban Mass Transportation Administration, *Transit Corridor Analysis—A Manual Sketch Planning Technique,* Washington, D.C., 1979.

7. U.S. Department of Transportation, Urban Mass Transportation Administration, Office of Transportation Economic Analysis, *Evaluating Transportation System Alternatives,* Washington, D.C., 1978.

8. Vuchic, Vukan R., *Urban Public Transportation, Systems and Technology,* Prentice-Hall, Englewood Cliffs, NJ, 1981.

9. U.S. Department of Transportation, Urban Mass Transportation Administration, *Characteristics of Urban Transportation Systems (CUTS),* Washington, D.C., 1978.

10. ———, 1985.

11. ———, 1992.

12. Institute of Transportation Engineers, *Transportation Planning Handbook,* Prentice-Hall, Englewood Cliffs, NJ, 1992.

13. Institute of Transportation Engineers, *Traffic Engineering Handbook,* Prentice-Hall, Englewood Cliffs, NJ, 1992.

14. Raus, J. *A Method for Estimating Fuel Consumption and Vehicle Emissions on Urban Arterials and Networks,* Report FHWA-TS-81-210, Office of Research and Development, Federal Highway Administration, Washington, D.C., 1981.

15. Voorhees, A.M. and Associates, Inc.: Handbook for Transportation System Management Planning, vol. 2, Prepared for the North Central Texas Council of Governments, August 1977.

Other relevant texts and documents are listed in the Bibliography.

4

Example: Service Policy Variables
and Design Projects

This hypothetical example of applying service policy variables to transportation planning and design is based upon illustrative physical and operational characteristics of a typical long-range corridor plan, to be completed at the sketch planning level. Based upon (1), and using the techniques and calculation methods of Chapters 2 and 3, the example demonstrates the mechanism for formulating the design for a given set of policy alternatives, conducting the demand analysis, preparing a preliminary physical design, and estimating the key impacts of capital cost, energy use, air pollution, and passenger travel time. It provides the basic information and the scope of work needed to develop and evaluate other alternatives that are the subject of the projects for solution, described at the end of the chapter.

The increase in demand for the corridor's transportation could be any combination of increases in population, economic activity, car ownership, or incomes that affect travel in a corridor between downtown and outlying parts of an urban area. Although the characteristics of the corridor shown in this example imply a considerable amount of urban growth, the principles illustrated in addressing the associated need for transportation are applicable to other degrees of urban development.

Summary of Required Scope of Work

The work to be done for the project is summarized as follows:

A preliminary transportation plan is required for the North Corridor of the city of Botolphville, which has approximately 1.8 million inhabitants. The

---

* This chapter is set with a unique style to emphasize the practical nature of the material covered. They are typical project calculations done in a design office.

completed plan is to include a description of the major transportation facilities (limited access and arterial highways and public transportation routes with operating details), together with estimates of the capital cost, capital cost per passenger kilometer, energy use, air pollution, and passenger travel time. Demographic and other data are provided in the following sections.

Urban Area Characteristics

Geographic and Land-Use Characteristics

The city of Botolphville is situated in a mid-Atlantic state of the United States and has a base year metropolitan area population of approximately 1.8 million. Its physical geography is that of a relatively flat coastal plain, bounded on the south by the ocean and on the east and west by rolling terrain. The North Corridor of this area is expected to provide the greatest amount of growth in future years. The metropolitan area's economic base comprises a mixture of historically significant port activities, trading, and light industry associated with a limited amount of coastal shipping, an agricultural hinterland, and a burgeoning electronics and service sector.

Existing Transportation System and Subject Improvements

The existing major transportation facilities are a costal highway, Interstate 100, a north-south interstate highway, I-90, and the existing Routes 12, 21, and 22 arterial highways. These facilities, together with the traffic zone system and related information, are shown in Figure 4-1. Also shown is the location of the transportation corridor of interest in this project (North Corridor), extending from the northern boundary of the downtown a distance of 5 miles. *This is the limit of the corridor for which the analysis will be conducted in this project and for which the extent of impacts must be estimated.*

Service Policy Variables

The plan is to be developed for the service policy variables shown in Table 4-1. *For purposes of the illustrative example, it is assumed that the service policy variables of interest are those directly related to the service level or cost of using the system as perceived by the individual at the time of use.*

The service policies in this example could be considered as the "status quo" in that the design level of service for automobiles on freeways is LOS C, and the variables associated with use of the transit system have no special inducements to make transit travel attractive to users. For example, the total travel time by transit is twice that by automobile, and the fare only half of the amount of the out-of-

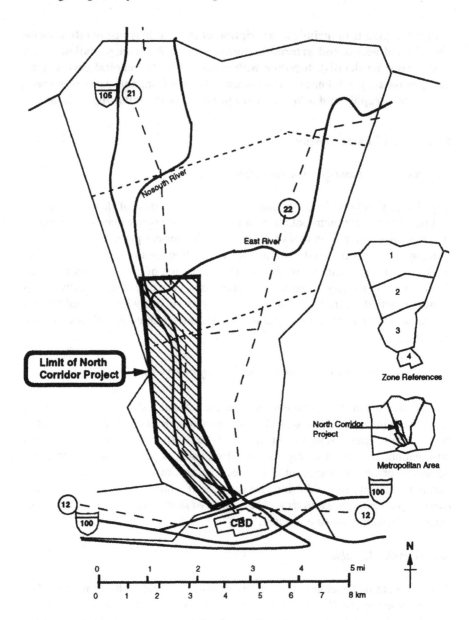

Figure 4-1   Study area characteristics

Table 4-1   Policy Variable Values for North Corridor Transportation Project

Zones	Mode	Service time, S [min]	Line-Haul time, L [min]	Out-of-pocket costs, C [cents]	Auto occ.	Auto LOS and pcphpl
ALTERNATIVE 1						
1–4:						
	Auto	5	30	600	1.2	C (1,200)
	Transit	25	50	100		
2–4:						
	Auto	5	28	580	1.2	C (1,200)
	Transit	20	42	85		
3–4:						
	Auto	5	12	550	1.2	C (1,200)
	Transit	20	22	75		

Note: Only zone pairs that include Zone 4 (predominantly the CBD) are expected to have significant public transportation connections. Therefore, only the policy variables associated with these zones are listed in this table.

pocket cost of driving a car. These are typical differences found in many North American cities in the mid 1990s. As will be seen in the final section of this chapter, this policy will form the basis for comparison with other policies that favor greater use of transit, with the objective of reducing adverse environmental impacts. The service policy alternatives specified here comprise the following variables for each of the zone pairs:

1. *Service time for transit users* (walking, waiting, transfer time).

2. *Cost differentials* in private vehicle use versus transit use, which may be varied by techniques such as parking surcharges and transit fare subsidies. Usually, cost differentials are expressed in terms of out-of-pocket costs to each user.

3. *Level of service (LOS) for automobile users.* Design LOS C is specified for freeways in the peak hours. Associated with this LOS are a travel speed and lane volume, which provide direct quantitative inputs to the analyses.

4. *In-vehicle travel time differentials* for the line-haul portion of the total trip for automobile versus transit.

5. *Vehicle occupancy for automobiles,* which, at 1.2 persons per car is average for most U.S. cities that do not experience significant carpooling for commuter trips.

Analysis Assumptions

For purposes of the analysis, it is assumed that inventories of land-use, demographic, and transportation features of the area have been completed and that projections of future population and other demographic characteristics have been made. It is also assumed that the demand analysis models have been satisfactorily calibrated to represent future conditions.

Analysis Format

To illustrate the context of the analysis and provide adequate detail to permit a realistic view of the procedures, each phase (described briefly in Chapters 2 and 3) is divided into the following parts: description, data, method, and output. An indication of the linkages between phases of the process is also given.

The assumptions stated in various parts of Chapters 2 and 3 about approximations in use of data and method, and the interpretation of results, are assumed to apply to the subject project also.

It should be noted that all input data and computations apply to the design year, unless stated otherwise. In many practical cases an analysis of the existing conditions would also be available. *However, we are assuming in this project that products of the study will apply to only the design year.*

The sequence of the project activities is shown in Figure 4-2.

Transportation Demand Analysis

The major basis for the analytical approach described in this section is the sequential transportation demand modeling process, the elements and selected methods of which were described in Chapter 2. The main elements following urban activity analysis are trip generation, trip distribution, modal split, and traffic assignment.

Urban Activity Analysis

Description

The urban area of interest was briefly described earlier. The available demographic data and zone referencing are described below.

Data

The demographic data for the design year are shown in Figure 4-3. Estimation of this population, income, and employment data is assumed here

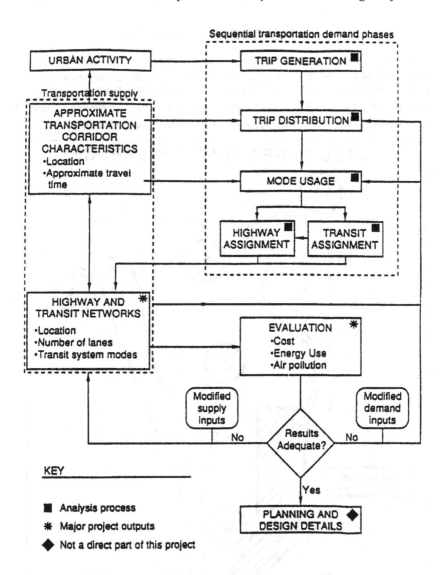

Figure 4-2 Flowchart of project activities

to have been determined from demographic records and economic projections.

The population data indicate that most residences are in Zone 1 and that there are relatively few residences in the city's fringe area and CBD (Zones 3 and 4, respectively).

The planning zones and their reference system, also shown in Figure 4-3, are as follows:

DEMOGRAPHIC DATA

Zone 1: Dwelling units - 50,000 Ave. income per DU = $32,000 Employees = 8,000
Zone 2: Dwelling units - 30,000 Ave. income per DU = $24,000 Employees = 8,000
Zone 3: Dwelling units - 20,000 Ave. income per DU = $16,000 Employees = 9,000
Zone 4: Dwelling units - 5,000 Ave. income per DU = $216,000 Employees = 30,000[1]

(1)   The 30,000 employees are those only from the zones shown in the study area.

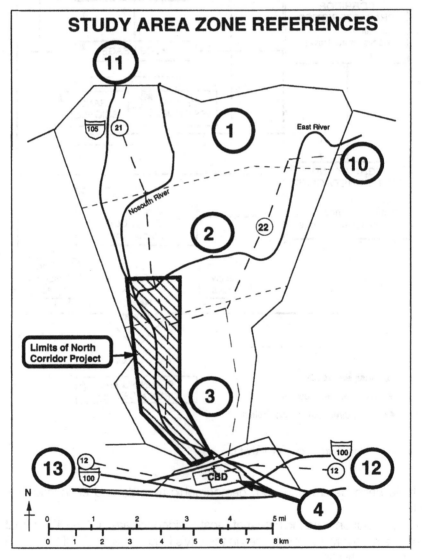

Figure 4-3   Demographic data and zone references

1. The internal zones are numbered 1 through 4. In practice, there would be more zones but each would be much smaller. The reduced number of zones used here permits a simplification in the computations while retaining the essence of the analysis process.

2. External zones are not shown, but the point at which each major transportation route crosses the study boundary is listed with a zone number. These are numbered 10 through 13, and represent the external zones. These zone numbers will be used in identifying internal-external trips and through trips at later stages of the analysis.

Linkages to Next Analysis Phase

The population levels and incomes listed in this phase will provide direct inputs to the trip generation phase of the demand estimation.

Trip Generation

Description

The three geographic categories of trip considered in this phase are internal trips, estimated by using household category and trip rate analysis, internal-external trips, and through trips. The internal trips will be described in some detail, and the latter two categories will be addressed separately, and later, in the trip assignment phase.

Data

Table 4-2 summarizes the demographic input data derived from the preceding phase, as well as trip attraction rates based upon an illustrative composite for all of the trip purposes.

Table 4-2   Demographic Data and Trip Attraction Rates

	TRIP GENERATION			
	Trip production		Trip attraction	
Zone	Number of DUs	Average income per DU	Number of employees	Trips per employee
1	50	32	8	16
2	30	24	8	12
3	20	16	9	3
4	5	16	30	12

Note: DUs, Incomes, and Employees are in 1,000s

The cross-classification curves relating income, automobile ownership, and trip production rates are shown in Figure 4-4, together with the resulting trip rates estimated for the project.

Method

Estimating productions and attractions

The process of estimating productions and attractions is shown in Table 4-3. The calculated attractions must be adjusted to equal the productions, as described in Chapter 2 and briefly noted below.

Balancing productions and attractions

In the example, because the trips are not broken down by purpose, the trip attractions for each zone have been adjusted to equal the trip productions for each zone in accordance with the formula:

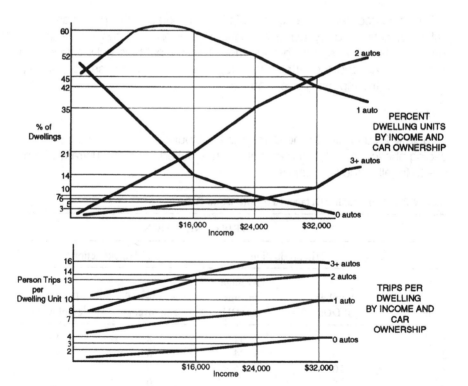

Figure 4-4   Cross-classification relationships for household trip generation analysis

Table 4-3   Internal Trip Generation Estimation

			Productions				Attractions			
Zone	Avg Income DUs Per DU	Autos Per DU	Percent DUs by Autos Owned	Total DUs by Autos Owned	Trips Per DU	Total Products P[i]	Employees	Trips per Employee	Total Attracts	Adjusted Total Attracts A[i]
1	50 32	0	3	1.50	4	6.00	8	16	128	231.52
		1	42	21.00	10	210.00				
		2	45	22.50	14	315.00				
		3+	10	5.00	16	80.00				
		Zone Totals	100	50.00		611.00				
2	30 24	0	7	2.10	3	6.30	8	12	96	173.64
		1	52	15.60	8	124.80				
		2	35	10.50	13	136.50				
		3+	6	1.80	16	28.80				
		Zone Totals	100	30.00		296.40				
3	20 16	0	14	2.80	2	5.60	9	3	27	48.84
		1	60	12.00	7	84.00				
		2	21	4.20	13	54.60				
		3+	5	1.00	14	14.00				
		Zone Totals	100	20.00		158.20				
4	5 16	0	14	0.70	2	1.40	30	12	360	651.15
		1	60	3.00	7	21.00				
		2	21	1.05	13	13.65				
		3+	5	0.25	14	3.50				
		Zone Totals	100	5.00		39.55				
Totals						1,105.15			611	1,105.15

Note: DUs, Income, Employees, Total Productions, and Total Attractions in 1,000s.

$$\text{Adjusted attr./zone} = \frac{(\text{computed attr./zone}) \ 3 \ (\text{total area prod./zone})}{(\text{total area attractions})}$$

Outputs

A trip end table showing the number of person productions and attractions for the internal trips for each zone for the 24-hour period of a typical weekday will be the output of this phase of the study. Table 4-4 provides a summary table of the results of this phase. These trip ends provide the inputs to the next phase, trip distribution.

A review of the outputs indicates that, as may be expected from the population and income levels for each zone, the highest trip productions occur in Zones 1, 2, and 3, in that order. Also, the greatest number of attractions occurs in Zone 4, the CBD, as would be expected. The total attractions initially estimated is considerably lower than the total productions (611,000 versus 1,105,150), and the balancing of the productions with the attractions is conducted in accordance with the formula above.

Table 4-4   Internal Trip Generation Estimation Summary

Zone No.	Productions	Attractions	
1	611.00	231.52	
2	296.40	173.64	◄── **Inputs to Trip**
3	158.20	48.84	**Distribution phase**
4	39.55	651.15	
Total	**1,105.15**	**1,150.15**	

Note: Productions, and Attractions in 1,000s.

Linkages to Next Analysis Phase

The productions and the adjusted attractions for each of the zones pro-vide direct inputs to the succeeding phase of the demand forecasting model, trip distribution.

Trip Distribution

Description

The gravity model is used here to estimate the 24-hour trip distribution, and then the peak hour distribution. This is then converted to a triangular matrix representing the numbers of person trips moving between the zones during the peak hour.

Data

The data for inputs to the trip distribution phase are the trip productions and attractions estimated in the previous phase, and the travel times and associated friction factors, summarized in Figure 4-5. These were obtained in a similar manner to that described in Chapter 2.

Method

The approach described in Chapter 2 for estimating the 24-hour trip inter-changes is shown in Table 4-5.

Outputs

The trip distribution known as the *trip interchange table* is summarized in Table 4-5 for the 24-hour trips, the design hour trips (K factor = 0.08), and as a triangular matrix indicating the number of trips moving *between* zones during the peak hour. The latter provide the inputs to the following phase of the demand modeling system, trip assignment. Also shown is the desire line

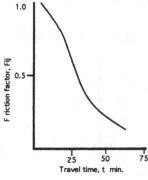

Travel Times [minutes]

From \ To	1	2	3	4
1	5	26	28	40
2	26	5	24	36
3	28	24	5	18
4	40	36	18	5

Friction Factors

From\ To	1	2	3	4
1	1.00	0.60	0.40	0.20
2	0.60	1.00	0.70	0.30
3	0.40	0.70	1.00	0.80
4	0.20	0.30	0.80	1.00

Figure 4-5   Travel time vs. friction factor

diagram, which indicates graphically the predominance of inbound trips to the CBD in the A.M. design hour, as well as a significant volume of trips on other routes.

Modal Split

Description

This phase first converts the peak hour person trip interchanges between zones estimated in the preceding trip distribution phase into the number of person trips for the automobile and the transit modes. Then these person trips are converted to vehicle trips, by means of an automobile occupancy factor, and to transit passenger trips for use in the succeeding phase, traffic assignment.

Data

The values of the major data inputs required to conduct the modal split analysis are listed in Table 4-6. They include:

1. The number of peak hour person trip interchanges between zones resulting from the trip distribution phase.
2. A list of the pairs of zones between which public transportation service will be offered. This reflects the fact that extensive transit service (apart from a limited amount of local service) may sometimes not be offered between certain pairs of zones where there is insufficient demand.

Table 4-5 Trip Distribution Estimation (24-Hr Passenger Trips in 1,000s)

	1	2	3	4	Sum Aij	Sum Aj Fij	Ri
Aj	231.52	173.64	48 84	651 15			

	Pi	tij	Fij									Sum Aij	Sum Aj Fij	Ri
1	611 00	5	1 0	26	0 6	28	0 4	40	0 2					
Iter 1, Aj*Fij		231.52		104.18		19.53		130.23					485.47	1.26
Iter 1, Pij		291.39		131 12		24.59		163.90				611.00		
Iter 2, Aj*Fij		138.02		71 45		17 32		207.59					434 38	1 41
Iter 2, Pij		194.15		100 50		24 36		292.00				611 00		

2	296.40	26	0.6	5	1 0	24	0 7	36	0 3				
		138.91		173 64		34.19		195.35				542 08	0.55
		75 95		94 94		18 69		106.81			296 40		
		82 81		119 08		30 30		311.38				543 58	0.55
		45 16		64.93		16 52		169 79			296 40		

3	158.20	28	0 4	24	0.7	5	1.0	18	0 8				
		92 61		121.55		48.84		520.92				783.91	0 20
		18.69		24 53		9.86		105 13			158.20		
		55 21		83.35		43 29		830.36				1,012 21	0 16
		8 63		13 03		6 77		129 78			158 20		

4	39.55	40	0.2	36	0.3	18	0 8	5	1 0				
		46.30		52 09		39.07		651.15				788.62	0.05
		2 32		2.61		1.96		32.66			39 55		
		27.60		35.72		34.63		1037.95				1,135.91	0.03
		0.96		1 24		1 21		36.14			39 55		

	1	2	3	4
Iter 1, Sum Pij	388.35	253 21	55 09	408.50
Target Aj + Current Aj	0 60	0.69	0 89	1.59
Iter 2, Sum Pij	248 89	179.70	48.85	627.71
Target Aj + Current Aj	0.93	0.97	1.00	1 04

24-Hour Passenger Trips (between zones) in 1,000s

	1	2	3	4
1	194 15	145 65	32.99	292.96
2		64 93	29.55	171.03
3			6.77	130.98
4				36 14

A.M. Peak Hour Person Trips (between zones) in 1,000s (K = 0.08)

	1	2	3	4
1	15.53	11.65	2.64	23.44
2		5.19	2.36	13.68
3			0.54	10.48
4				2.89

Inputs to modal split phase

Zone 1
Zone 2
Zone 3
Zone 4
Desire line diagram

3. The calibrated utility equations incorporating the appropriate policy variables specified in the Policy Summary presented earlier. The variables include:

    (a) Estimated service time (walking and waiting time) between zones for automobile and public transportation.

    (b) Estimated line-haul, or in-vehicle travel times, between zones for automobile and public transportation modes

    (c) Estimated out-of-pocket travel costs between zones for automobile and public transportation modes.

Table 4-6 Modal Split Inputs

1. Outputs from Trip Distribution:
A.M. Peak Hour Person Trips (between zones) in 1,000s (K = 0.08)

	1	2	3	4
1	15.53	11.65	2.64	23.44
2		5.19	2.36	13.68
3			0.54	10.48
4				2.89

2. Binomial Logit Model

$$p(k) = \frac{e^{U_k}}{\sum\limits_{x=1} e^{U_x}}$$

where:

K = mode
$p(K)$ = proportion of trips that will select mode K
U_K = utility function of mode K
U_x = utility function of mode x

Utility functions
U_K = $a(K) - 0.07S - 0.015L - 0.0020C$ (Zones 1 to 4)
U_K = $a(K) - 0.07S - 0.015L - 0.0025C$ (Zones 2 to 4)
U_K = $a(K) - 0.07S - 0.015L - 0.0030C$ (Zones 3 to 4)

where:

$a(K)$ = precalibrated constant for mode K
     = -0.8 for autos
     = -0.1 for transit
S = service time [minutes]
L = line-haul time [minutes]
C = out-of-pocket costs [cents]

3. Zone Pairs with Transit Service

Zones 1 to 4
Zones 2 to 4
Zones 3 to 4

Table 4-7   Modal Split Estimation

Zones	Mode	S	L	C	U[K]	Exp(U[K])	P[K]
1–4:	Auto	5	30	600	−2.80	0.06	0.50
	Transit	25	50	100	−2.80	0.06	0.50
					Sum	0.12	1.00
2–4:	Auto	5	28	580	−3.02	0.05	0.34
	Transit	20	42	85	−2.34	0.10	0.66
					Sum	0.14	1.00
3–4:	Auto	5	12	550	−2.98	0.05	0.28
	Transit	20	22	75	−2.06	0.13	0.72
					Sum	0.18	1.00

	SUMMARY AM PEAK HOUR TRIPS			COMMENTS
Zones	1 to 4	2 to 4	3 to 4	
Total Pass-Trips	23.44	13.68	10.48	From trip distribution
% Transit-Trips	50%	66%	72%	See table above
Transit Pass-Trips	11.72	*9.07*	*7.50*	Total pass trips × % transit trips
Auto Pass-Trips	11.72	4.61	2.98	Total pass trips − Transit pass trips
Auto Occupancy	1.20	1.20	1.20	Specified policy variable
Auto Vehicle-Trips	*9.77*	*3.84*	*2.48*	Auto pass trips/Auto occ.

Notes: All trips in 1,000s.

Only Zones 1–4, 2–4, and 3–4 are included here, because transit between other zone pairs is insignificant.

The values of S, L, and C are based upon the relevant policy variables stated earlier.

Transit pax trips and auto vehicle trips shown thus: *11.72*   etc.

are inputs to the next phase, trip assignment

4. The vehicle occupancy values in the Policy Summary used to con-
    vert person trips to vehicle trips.

Method

The logit model modal split analysis provides an estimate of the proportion of persons using public transportation between any two zones, based

upon the estimated utilities of each mode for users in each zone. The analysis and results are shown in Table 4-7.

The person trips by automobile are converted to vehicle trips by dividing person trips by the average vehicle occupancy. Person trips by public transport are left as person trips for input to the next phase—traffic assignment. These public transit person trips will not be converted to vehicle trips until after the trip assignment phase because, until the link volumes are known, the appropriate transit mode and, hence, the vehicle occupancy, cannot be determined.

Outputs

The transit passenger trips and the auto vehicle trips during the A.M. peak hour are the major outputs.

Key points of the results indicate that:

1. The probability of a trip being made by transit increases with the proximity of the zone to the CBD. This is consistent with typical experience where travelers from city fringe areas or inner suburbs are most likely to use transit for trips to the CBD. For Zone 3, for example, the modal split is 72% transit, whereas for Zone 1 the corresponding probability is 50%.

2. When the modal split probability is combined with the trip productions, the greatest number of trips are from Zone 1, and the least number from Zone 3.

Linkages to Following Phase

These values estimated will be used as the major inputs to the next phase, traffic assignment.

Traffic Assignment

Description

In this phase the automobile vehicle volume and the transit passenger volume between each of the zone pairs from the preceding modal split phase are assigned to the highway and transit network. The automobile and transit passenger volumes for the "design links" within the corridor of interest are then identified to provide a basis for estimating the extent of the roads and transit facilities in the following phase, physical facilities design.

Data

The following items of data are required:

1. The automobile volume and transit passenger volume moving between each of the zones in the study area are shown in Table 4-8. These data are obtained from the foregoing modal split phase for the trips on the transit routes. For the trips on the nontransit routes, the number of auto trips is obtained from the trip distribution phase, by dividing the relevant trips by the vehicle occupancy.

2. The trip end matrices for the internal-external and the through traffic movements are also shown in Table 4-8. These movements are all assumed to be vehicles moving on the highway system and do not include transit vehicles, except for intercity buses that are insignificant in terms of the total traffic movement. The volumes listed in these matrices are assumed to have been obtained from other regional traffic studies independently of the subject project.

3. The coded network diagrams for automobile and transit are shown in Figure 4-6. The link travel times are also shown for the entire corridor and for the portions of the network that lie within the project limits. This network, together with the trip interchanges listed in Table 4-8, provides the basis for identifying the minimum time paths and the volumes on each link.

Table 4-8 Trip Assignment Inputs

Internal Auto Trips
(A.M. Peak Hour Vehicle Trips in 1,000s
Auto Occupancy = 1.20)

	1	2	3	4
1		9.71	2.20	9.77
2			1.97	3.84
3				2.48
4				

Through Auto Trips
(A.M. Peak Hour Vehicle Trips in 1,000s)

	10	11	12	13
10		0.2	0.3	0.6
11			0.2	0.1
12				0.5
13				

Internal-External Auto Trips
(A.M. Peak Hour Vehicle Trips in 1,000s)

	10	11	12	13
1	0.1	0.1	0.1	0.1
2	0.2	0.2	0.2	0.2
3	0.3	0.3	0.3	0.3
4	0.4	0.4	0.4	0.4

Transit Trips
(A.M. Peak Hour Passenger Trips in 1,000s)

	1	2	3	4
1				11.72
2				9.07
3				7.50
4				

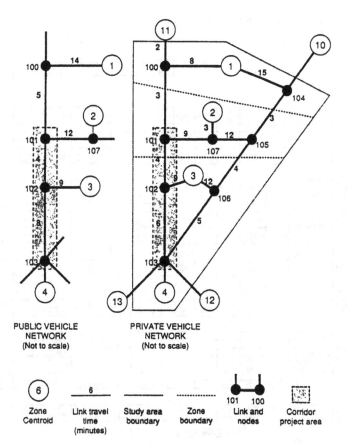

Figure 4-6   Trip assignment network diagram

Method

For this preliminary analysis, the trip assignment method will be the "all or nothing" process using the design hourly volumes (in this case the A.M. inbound peak hour of traffic) on the least time paths between origins and destinations. The peak hour volumes will be modified by the directional factor (D) to provide directional design hourly volumes (DDHVs), which will form the basis for facilities planning. The format for assembling the travel times by links for each of the trip pairs for the internal, internal-external, and through tips for automobiles and transit passengers, and for computing the volumes on each link, is shown in Table 4-9. Note that although the link volumes are computed for each of the links in the network, only the volumes on those links within the project area are used for the facilities design described in the ensuing sections.

Table 4-9   Trip Assignment Estimates

A.M peak hour trips in 1,000s

Orig Dest	Trips	11 100	100 101	101 102	102 103	1 100	2 107	3 102	10 104	101 107	104 105	105 106	105 107	106 103	1 104	3 106	103 13	103 4	103 12
PRIVATE VEHICLE TRIPS																			
Internal																			
1-2	9 71		9.71			9.71	9.71			9.71									
1-3	2.20		2.20	2.20		2.20		2.20											
1-4	9 77		9.77	9.77	9.77	9.77											9.77		
2-3	1.97			1.97			1.97	1.97		1.97									
2-4	3.84			3.84	3.84		3.84			3.84								3.84	
3-4	2.48				2.48			2.48										2.48	
Total	29.97	0.00	21.68	17.78	16.09	21.68	15 52	6.65	0 00	15.52	0.00	0.00	0.00	0.00	0.00	0 00	0.00	16.09	0 00
Internal-External																			
1-10	0.10								0.10					0 10					
1-11	0.10	0.10				0.10													
1-12	0.10		0 10	0 10	0.10	0.10													0.10
1-13	0.10		0.10	0 10	0.10	0 10									0.10				
2-10	0.20						0.20		0.20		0.20		0.20						
2-11	0.20	0.20	0.20				0.20			0.20									
2-12	0 20			0.20	0 20		0.20			0.20									0 20
2-13	0.20			0 20	0.20		0.20			0.20						0.20			
3-10	0.30								0.30		0.30	0.30				0.30			
3-11	0.30	0.30	0.30	0 30				0.30											
3-12	0.30				0 30			0 30											0.30
3-13	0 30			0.30				0.30							0.30			
4-10	0.40								0.40		0.40	0 40		0.40				0.40	
4-11	0.40	0.40	0.40	0.40	0 40													0.40	
4-12	0.40																	0.40	0.40
4-13	0.40														0.40	0.40			
Total	4.00	1.00	1.10	1.30	1 60	0.30	0.80	0.90	1 00	0.60	0.90	0.70	0.20	0.40	0.10	0 30	1.00	1.60	1.00
Through																			
10-11	0.20	0.20				0.20			0.20					0.20					
10-12	0.30								0.30		0.30	0.30		0.30					0.30
10-13	0 60								0.60		0.60	0.60		0 60		0.60			
11-12	0.20	0.20	0 20	0 20	0.20														0.20
11-13	0.10	0.10	0.10	0.10	0.10										0.10			
12-13	0.50														0.50	0.50		
Total	1.90	0.50	0.30	0.30	0.30	0.20	0.00	0 00	1 10	0.00	0.90	0.90	0 00	0 90	0.20	0,00	1 20	0.00	1.00
Totals																			
2-Way	23.32	1.50	23.08	19 38	17.99	22.18	16.32	7.55	2 10	16.12	1.80	1.60	0.20	1.30	0.30	0.30	2.20	17.69	2.00
1-Way	0.98	0.98	15 00	12.60	11.69	14.42	10.61	4.91	1.37	10.48	1.17	1.04	0.13	0.85	0.20	0.20	1.43	11.50	1 30

D = 0.65

|Input to Facilities Design|

TRANSIT PASSENGER TRIPS

Internal																			
1-4	11.72		11.72	11 72	11.72	11 72												11.72	
2-4	9 07			9.07	9.07		9 07			9.07								9.07	
3-4	7.50				7.50			7.50										7.50	
2-Way	28 29	0.00	11.72	20.79	28.29	11.72	9.07	7.50	0.00	9.07	0.00	0.00	0.00	0.00	0.00	0.00	0.00	28.29	0.00
1-Way	18.39	0.00	7 62	13.51	18.39	7.62	5 90	4.88	0.00	5.90	0 00	0.00	0.00	0 00	0.00	0.00	0.00	18.39	0.00

D = 0.65

|Input to Facilities Design|

Outputs

The outputs of this phase are the automobile design hour volumes and transit passenger design hour volumes by predominant direction for each link in the network. By inspection, the link for which the corridor volumes must be estimated is link number 102-103. The estimated volumes, i.e., 11,690 automobiles per hour and 18,390 transit passengers per hour, are then used directly in estimating the physical and operational facilities, described in the next phase. Note that certain other links in the network have higher volumes of automobile traffic than link 102-103. This may well occur in practice, but our objective is to design the facilities for link 102-103 only; thus, we will concentrate our efforts solely on this link in the ensuing design phases.

Notes and Comments

In this example, the number of zones and the extent of the transportation network have been kept to a minimum in order to simplify the computations and make them amenable to the use of a pocket calculator or computerized spreadsheet. If the travel time on any link were to be altered to the extent that a different route between two nodes occurred, the table would have to be rearranged accordingly.

Physical Facilities Design

Description

In this phase the number of highway lanes or rail tracks needed to accommodate the demand on the design link of the corridor is estimated. This enables the cross-sectional dimensions of the facilities to be determined as a basis for the cost estimate and for exploring possible detailed route locations within the urban area. When these and the operating features of the system have been decided, the impacts can be estimated.

Data

The data items needed for this stage, described in greater detail in Chapter 3, are listed in Table 4-10, along with the values defined in the specified policy or assumed for default purposes for this preliminary design. As indicated earlier, the modes to be considered include automobiles and buses on existing arterials and freeways, and BRT, LRT, and RRT.

Method

The approach is similar to that shown in Chapter 3, and is detailed in Table 4-11. An assumption is made about the volumes of bus passengers and automobiles that will be accommodated on upgraded arterial highways. Two options, A and B, are then examined. Option A uses BRT and LRT for transit passengers; Option B uses only RRT.

Outputs

The number of lanes for automobiles and buses and the number of tracks for LRT in the peak hour are the major outputs. They show that for both

Table 4-10   1-Way Lane or Track Requirements: Link 102–103, Inputs

Facility	Item	Units	Range of values	Comments
Freeways, automobiles	Service volume	Pcphpl	1,200	LOS C
Freeways, bus, no priority	Bus pass volume	Pass/hr	1,000	Assumed
	Bus occupancy	Pass/bus	50	Assumed
	Bus pce	Number	2	Default, Ref. (1)
Bus rapid transit (BRT)	Bus occupancy	Pass/bus	50	Assumed
	Service volume	Bus/hr/la	120–180	20 sec min. practical headway
	Pass lane vol	Pass/la/hr	6,000–12,000	Service vol. * occupancy
Light rail transit (LRT)	Train occupancy	Pass/train	300	2 cars/train * 150 pax/car
	Service volume	Train/hr/tr	20–30	120 sec min. practical headway
	Pass track vol.	Pass/la/hr	6,000–9,000	Service vol. * occupancy
Rail rapid transit (RRT)	Train occupancy	Pass/train	900	6 cars/train * 150 pax/car
	Service volume	Train/hr/tr	15–30	120 sec min. practical headway
	Pass track vol.	Pass/la/hr	13500–27000	Service vol. * occupancy

options eight lanes of freeway (one-way)—most likely in two separate facilities—are required for automobile traffic. For Option A one BRT lane and one LRT track are required. For Option B one RRT route satisfies the entire transit passenger demand.

Impacts

The impacts to be investigated are:

* Capital cost and capital cost per passenger kilometer
* Energy consumption
* Air pollution (carbon monoxide).

Because at this stage two facility design options are under consideration (Options A and B), and because the main objective is to illustrate the procedure of estimating the impacts, we will use the capital cost per passenger kilometer and other operational considerations to eliminate one option from further consideration. The procedure for estimating the impacts if other options were also being considered would be essentially similar. A simplified

Table 4-11   1-Way Lane or Track Requirements: Link 102–103

ITEM	UNITS	OPTION A	OPTION B	COMMENTS
DESIGN YEAR VOLUMES (From trip assignment phase)				
Transit Pass Volume	Pass / hr	18,390	18,390	From Trip assignment total 1-way Link 102-103
Auto Volume	Pcph	11,690	11,690	From Trip assignment total 1-way Link 102-103
DEMAND ABSORBED BY UPGRADED SURFACE ARTERIALS (NON FREEWAY) IN THE DESIGN YEAR				
Bus Pass Volume	Pass / hr	2,200	2,200	Assumed values for illustrative purposes
Auto Volume	Pcph	3,600	3,600	Assumed values for illustrative purposes
ADJUSTED DESIGN YEAR VOLUMES FOR FREEWAY AND RAIL FACILITIES				
Transit Pass Volume	Pass / hr	16,190	16,190	Previous Transit Pass Volume - Arterial Bus Pass Vol.
Auto Volume	Veh / hr	8,090	8,090	Previous Auto Volume - Arterial Auto Volume

New Freeways and High Capacity Transit

ITEM	UNITS	OPTION A	OPTION B	COMMENTS
HIGHWAY VEHICLES (AUTOMOBILES) (1)				
Pce Volume	Pceph	8,090	8,090	Auto Volume
Service Volume	Pcephpl	1,200	1,200	LOS C = 1,200, LOS E = 2,000 pcphpl
Lanes Required	Number	6.74	6.74	Pce Volume / Service Volume
		7	7	Rounded
ADJUSTED DESIGN YEAR VOLUMES				
Transit Pass Volume	Pass / hr	16,190	16,190	Transit Pass Volume (1)
Auto Volume	Veh / hr	0	0	Previous Auto Volume - Highway Auto Volume
BUS RAPID TRANSIT (BRT)				
BRT Pass Volume	Pass / hr	8,000	0	Assumed
Bus Occupancy	Pass / bus	50	50	Assumed
Bus Volume	Bus / hr	160	0	BRT Pass Volume / Bus Occupancy
Service Volume	Bus / hr / la	180	180	20 sec min. practical headway
Lanes Required	Number	0.89	0.00	Bus Volume / Service Volume
		1	0	Rounded
ADJUSTED DESIGN YEAR VOLUMES				
Transit Pass Volume	pax / hr	8,190	16,190	Previous Transit Pass Volume - BRT Pass Volume
LIGHT RAIL TRANSIT (LRT)				
LRT Pass Volume	Pass / hr	8,190	0	LRT matches remaining passenger volume for Option A
Train Occupancy	Pass / train	300	300	2 cars / train * 150 pass / car
Train Volume	Train / hr	27	0	LRT Pass Volume / Train Occupancy
Service Volume	Train / hr / tr	30	30	120 sec minimum practical headway
Tracks Required	Number	0.91	0.00	Train Volume / Service Volume
		1	0	Rounded
ADJUSTED DESIGN YEAR VOLUMES				
Transit Pass Volume	Pass / hr	0	16,190	Previous Transit Pass Volume - LRT Pass Volume
RAPID RAIL TRANSIT (RRT)				
RRT Pass Volume	Pass / hr	0	16,190	RRT matches remaining passenger volume for Option B
Train Occupancy	Pass / train	900	900	6 cars / train * 150 pass / car
Train Volume	Train / hr	0	18	RRT Pass Volume + Train Occupancy
Service Volume	Train / hr / tr	30	30	120 sec minimum practical headway
Tracks Required	Number	0	0.60	Train Volume + Service Volume
		0	1	Rounded
ADJUSTED DESIGN YEAR VOLUMES				
Transit Pass Volume	Pass / hr	0	0	Previous Transit Pass Volume - RRT Pass Volume

(1) Assumes all highway vehicles are automobiles, although a small percentage of buses (no priority) may be present

flowchart describing the process of selecting one of the options is shown in Figure 4-7.

Capital Cost and Cost per Passenger Kilometer

Description

In this phase we estimate the capital cost and the capital cost per passenger kilometer of Options A and B formulated in the preceding phase. The least cost per passenger kilometer and other features of the options provide a basis for selecting one of them for investigation of the air pollution and energy consumption.

The extent of the fixed facilities estimated in the facilities design, plus the cost of stations and the transit vehicle fleets, will result in the initial, or capital, cost of the proposed project.

Data

The unit costs used are shown in Table 4-12. No costs for land acquisition are included because these would be unique to each location.

Method

A tabular approach as shown in Chapter 3 is adopted. The capital costs for each alternative are shown in Table 4-13, along with the capital costs per

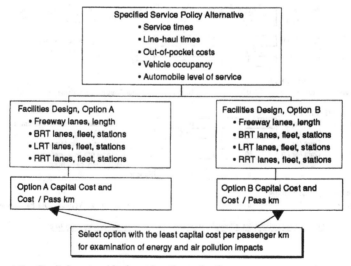

Figure 4-7 Capital cost analysis and option selection

Table 4-12   Capital Cost Inputs

Facility		Item	Units	Value	Comments
Freeways		Length	km	8	Specified project length
		Unit cost	$/lane km	1,366,200	Table 3-1
BRT	**Way:**	Length	km	8	Specified project length
		Unit cost	$/lane km	1,366,200	Table 3-1
	Fleet:	Ave Speed	km/h	40	Selected for illustrative example
		Unit cost	$/bus	190,000	Table 3-1
	Stations:	Stations	Number	2	Selected for illustrative example
		Unit cost	$/station	1,000,000	Table 3-1
LRT	**Way:**	Length	km	8	Specified project length
		Unit cost	Route km	12,420,000	Table 3-1 (includes stations)
	Fleet:	Ave Speed	km/h	40	Selected for illustrative example
		Unit cost	$/car	1,600,000	Table 3-1
	Stations:	Stations	Number	3	Selected for illustrative example
		Unit cost	$/station	1,000,000	Table 3-1
RRT	**Way:**	Length	km	8	Specified project length
		Unit cost	Route km	43,470,000	Table 3-1 (includes stations)
	Fleet:	Ave Speed	km/h	48	Selected for illustrative example
		Unit cost	$/car	1,200,000	Table 3-1
	Stations:	Stations	Number	2	Selected for illustrative example
		Unit cost	$/station	4,500,000	Table 3-1

passenger kilometer. Note that estimates of the fleet sizes for BRT, LRT, and RRT must be made, and the lane and track costs are for two-way facilities, expressed as "route km." The unit costs reflect this.

Selection of Option

Examination of the results of the capital cost analysis shows that the capital costs of Options A and B are approximately $482 million and $728 million, respectively. The costs per passenger kilometer are $18,597 and $27,135 respectively. Clearly, Option A is the least expensive in terms of capital cost.

An additional consideration to the above costs is that Option A would provide some redundancy to the system, i.e., if the LRT system were to experience a power failure or other service interruptions, the bus system would be available as a backup to provide emergency service. *Therefore, a decision is made at this point in this idealized example to proceed with Option A for investigation of the energy consumption and air pollution impacts.*

Table 4-13 Capital Cost Estimates—Total and per Passenger Kilometer

ITEM	UNITS	$ CAPITAL COST OPTION A	OPTION B	COMMENTS	PASSENGERS OPTION A	OPTION B	$ COST / PASS KM. OPTION A	OPTION B
FREEWAYS (Automobiles)								
Length	Km	8	8	Project length				
Lanes	Number	7	7	From Lane/Track Requirements, Table 4-11				
Lane km	Lane km	112	112	Length x Lanes x 2-way				
Unit cost	$ / lane km	1,366,200	1,366,200	From Table 4-12				
Cost	$	306,028,800	306,028,800	Lane km x Unit Cost	9,708	9,708	31,523	31,523
BRT								
Way:								
Length	Km	8	0	Project length				
Lanes	Number	1	0	From Lane / Track Requirements				
Lane km	Lane km	16	0	Length x Lanes x 2-way				
Unit cost	$ / lane km	1,366,200	1,366,200	Table 4-12				
Cost	$	21,859,200	0	Lane Miles x Unit Cost				
Fleet:								
Frequency	Bus / hr	160	0	From Lane / Track Requirements				
Average Speed	Km/h	48	48	Assumed				
Distance	km	16	0	2 x Length (round trip)				
Cycle Time	Hours	0.33	0.00	Distance / Average Speed				
Fleet	Number	53	0	Frequency x Cycle Time				
Unit cost	$ / bus	190,000	190,000	Table 4-12				
Cost	$	10,133,333	0	Fleet x Unit Cost				
Stations:								
Stations	Number	3	0	Assumed				
Unit cost	$ / station	1,000,000	1,000,000	Assume = LRT station cost				
Cost	$	3,000,000	0	Stations				
TOTAL, BRT		34,992,533	0		8,000	0	4.37	#DIV/0!
LRT								
Way:								
Length	Km	8	0	Project length				
Routes	Number	1	1	From Lane / Track Requirements				
Distance	Route km	8	0	Length x tracks				
Unit cost	$ / route km	12,420,000	12,420,000	From Table 4-12				
Cost	$	99,360,000	0	Route km Unit Cost				
Fleet:								
Frequency	Trains / hr	24	0	From Lane / Track Requirements				
Average Speed	Km/h	30	30	Assumed				
Distance	Km	16	0	2 x Length (round trip)				
Cycle Time	Hours	0.53	0.00	Distance / Average Speed				
Fleet	Number	13	0	Frequency x Cycle Time				
Unit cost	$ / train	3,222,000	3,222,000	Table 4-12 x cars/train				
Cost	$	41,241,600	0	Fleet x Unit Cost				
Stations:								
Included in way costs								
TOTAL, LRT		140,601,600	0		8,190	0	17,167	#DIV/0!
RRT								
Way:								
Length	Km	0	8	Project length				
Routes	Number	0	1	From Lane / Track Requirements				
Distance	Route km	0	8	Length x Tracks				
Unit Cost	$ / track km	0	43,470,000	Table 4-12				
Cost	$	0	347,760,000	Route km x Unit Cost				
Fleet:								
Frequency	Trains / hr	0	17	From Lane / Track Requirements				
Average Speed	Km/h	50	50	Assumed				
Distance	Route km	0	16	2 x Length (round trip)				
Cycle Time	Hours	0.00	0.32	Distance / Average Speed				
Fleet	Number	0	5.44	Frequency x Cycle Time				
Unit Cost	$ / train	9,000,000	9,000,000	Table 4-12 x cars/train				
Cost	$	0	48,960,000	Fleet x Unit Cost				
Stations:								
Included in way costs								
TOTAL, RRT		0	396,720,000		0	16,190	#DIV/0!	24,504
TOTAL PROJECT		481,622,933	702,748,800		25,898	25,898	18,597	27,135

Outputs

The major outputs are the capital cost and the capital cost per passenger kilometer. These values will be shown in the impact summary sheet to assist in the evaluation of the project.

Energy Requirements

Description

The amount of energy needed to operate the system during the peak hour will result from the sum of the energy requirements of each mode. In this analysis we are not including the energy needed during construction, and this may be considerable, especially if significant portions of the routes require tunneling, bridges, or other high-cost items.

Data and Method

The items needed for the energy estimate were described earlier in Chapter 3 and are listed in Table 4-14. The energy usage is estimated in the tabular format in Table 4-15. Again, subtotals are shown for each of the modes.

Table 4-14 Energy Consumption Estimation Inputs

Item	Units	Value	Comments
Autos on limited access highways			
Length	km	8	Specified project length
Energy consumption factor	MJ/auto km	4	From Table 3-3
BRT			
Length	km	8	Specified project length
Energy consumption factor	MJ/Bus km	26	From Table 3-3
LRT			
Length	km	8	Specified project length
Energy consumption factor	MJ/LRT veh (car) km	32	From Table 3-3
RRT			
Length	km	8	Specified project length
Energy consumption factor	MJ/RRT veh (car) km	40	From Table 3-3

Table 4-15   Energy Consumption Estimate [For design hour volume (DHV)]

ITEM	UNITS	VALUE	COMMENTS
AUTOS ON LIMITED ACCESS HIGHWAYS			
Length	km	8	Specified project length
DHV	pcph	12,446	DDHV (from Table 4-11) / 0.65 (2-way traffic)
Distance traveled	veh km	99,569	Length x DHV
Gasoline energy	MJ / veh km	4.00	Table 4-12, includes some trucks
Energy use	MJ	398,277	Distance traveled x gasoline energy
BRT			
Length	km	8	Specified project length
DHV	veh / hr	320	DDHV (From Table 4-11) x 2
Distance traveled	veh km	2,560	Length x DHV
Diesel energy	MJ / veh km	26.00	From Table 4-12
Energy use	MJ	66,560	Distance traveled x diesel energy
LRT			
Length	km	8	Specified project length
DHV	train / hr	54	DDHV (From Table 4-11) x 2-way *
Distance traveled	veh km	864	Length x DHV x 2 cars/train
Oil energy	MJ / veh km	32.00	From Table 4-12
Energy use	MJ	27,648	Distance traveled x oil energy
RRT			
Length	km	8	Specified project length
DHV	veh / hr	0	DDHV (From Table 4-11) x 2-way
Distance traveled	veh km	0	Length x DHV x 6 cars/train
Oil energy	MJ / veh km	37.00	From Table 4-12
Energy use	MJ	0	Distance traveled x oil energy
TOTAL ENERGY MJ		**492,485**	

Outputs

The major output is the total amount of energy consumed during the peak hour of travel. It amounts to approximately 492,000 MJ. This amount will be shown later in the impact summary sheet to assist in the evaluation of the project.

Air Pollution

Description

The amount of air pollution only in terms of carbon monoxide (CO) emissions during the peak hour is estimated here. The amounts for each mode are computed and summed to obtain a total for the entire project.

Data and Method

The items needed for the CO estimate were described earlier in Chapter 3, and the unit values are listed in Table 4-16. Estimation of the total CO emissions is conducted in the tabular format shown in Table 4-17.

Outputs

The major output is the total amount of CO emitted during the peak hour of travel. For this project it amounts to approximately 1,048 kg. This amount will be shown later in the impact summary sheet to assist in the evaluation of the project.

Summary of Findings

Numerical Results

The summarized results of the project for this policy alternative are shown in Table 4-18. The key findings are discussed below.

Table 4-16   CO Air Pollution Inputs

Item	Units	Value	Comments (1)
Autos on limited access highways			
Length	km	8	Specified project length
CO Emissions Factor	gm/veh km	9.00	Table 3-4, LOS C
BRT			
Length	km	8	Specified project length
CO Emissions Factor		10.19	Table 3-4, LOS C
LRT			
Length	km	8	Specified project length
CO Emissions Factor	gm/LRT veh (car) km	0.0062	Table 3-4
RRT			
Length	km	8	Specified project length
CO Emissions Factor	gm/RRT veh (car) km	0.0062	Table 3-4

(1) LOS C for a 60 mph [97 km/h] free-flow freeway occurs at 60 mph [97 km/h] (Ref. (2)).

LOS E for a 60 mph [97 km/h] free-flow freeway occurs at 50 mph [80 km/h] (Ref. (2)).

Table 4-17   CO Air Pollution Estimates (For design hour volume [DHV])

ITEM	UNITS	VALUE	COMMENTS
AUTOS ON LIMITED ACCESS HIGHWAYS			
Length	km	8	Specified project length
DHV	pcph	14,125	DDHV (from Table 4-11) ÷ 0.65 (2-way traffic)
Distance traveled	veh km	113,000	Length x DHV
CO Emissions Factor	gm/veh km	9.00	From Table 4-16, LOS C
CO Emissions	gm	1,017,000	Distance traveled x CO Emissions Factor
BRT			
Length	km	8	Specified project length
DHV	pcph	360	DDHV (from Table 4-11) x 2
Distance traveled	veh km	2,880	Length x DHV
CO Emissions Factor	gm/veh km	10.19	From Table 4-16, LOS C
CO Emissions	gm	29.347	Distance traveled CO Emissions Factor
LRT			
Length	km	8	Specified project length
DHV	pcph	54	DDHV (from Table 4-11) x 2
Distance traveled	veh km	864	Length x DHV x cars/train
CO Emissions Factor	gm/veh km	0.01	From Table 4-16
CO Emissions	gm	9	LRT car (veh) km x CO Emissions Factor
RRT			
Length	km	0	Specified project length
DHV	pcph	0	DDHV (From Table 4-11) x 2
Distance traveled	veh km	0	Length x DHV x cars/train
CO Emissions Factor	gm/veh km	0.01	From Table 4-16
CO Emissions	gm	0	RRT car (veh) km x CO Emissions Factor
TOTAL CO EMISSIONS	**kg**	**1,017**	

Capital Cost

The total capital cost of the project is over $482 million. Of this, 64% is for highway construction, while the proportion of highway users is 37%. This apparently disproportionate cost per passenger is confirmed when it is seen that the cost on a per passenger basis is more than $31,000. In contrast, bus and rapid transit facilities require approximately $4,000 and $17,100 per passenger, respectively.

Energy Use

Of the total of over 492,000 MJ expended on energy, 81% is used for automobiles. This is approximately 41 MJ per passenger for the automobile mode. This compares with approximately 11 MJ per passenger for all of the other modes combined. On this basis, it is clear that the average automobile passenger accounts for about four times the amount of energy used by an average bus or rail transit passenger. Of course, it must be realized that this applies during the peak hour, when buses are assumed to carry 50 pas-

Table 4-18 Summary of Costs, Energy, and Pollution

ITEM	CAPITAL COST		PASSENGERS		AVE. CAPITAL COST ($) PER PASSENGER
	AMOUNT ($)	PERCENT	NUMBER	PERCENT	
FREEWAY:					
Subtotal:	306,028,800	64%	9,708	37%	31,523
BRT:					
Way	21,859,200				
Fleet	10,133,333				
Station	3,000,000				
Subtotal	34,992,533	7%	8,000	31%	4,374
LRT:					
Way	99,360,000				
Fleet	41,241,600				
Subtotal	140,601,600	29%	8,190	32%	17,167
RRT:					
Way	0				
Fleet	0				
Subtotal	0	0%	0	0%	#DIV/0!
TOTAL	481,622,933	100%	25,898	100%	18,597

ENERGY

ITEM	ENERGY USE		PASSENGERS		AVE. ENERGY USE (MJ) PER PASSENGER
	AMOUNT (MJ)	PERCENT	NUMBER	PERCENT	
Automobiles	398,277	81%	9,708	37%	41
BRT	66,560	13%	8,000	31%	8
LRT	27,648	6%	8,190	32%	3
RRT	0	0%	0	0%	0
TOTAL	492,485	100%	25,898	100%	19

CO EMISSIONS

ITEM	EMISSIONS		PASSENGERS		AVE. CO EMISSIONS (GM) PER PASSENGER
	AMOUNT (gm)	PERCENT	NUMBER	PERCENT	
Automobiles	1,017,000	97%	9,708	37%	105
BRT	29.347	0%	8,000	31%	0
LRT	9	0%	8,190	32%	0
RRT	0	0%	0	0%	0
TOTAL	1,046,356	100%	25,898	100%	40

sengers per bus and rail transit vehicles are also running at near capacity. For a considerable portion of the day during the off-peak hours, transit vehicles will be operating with many fewer passengers. Consequently, their energy efficiency in terms of passengers per unit of energy consumed on a daily basis will be considerably lower than for peak hour operations.

Air Pollution—Carbon Monoxide

The total amount of CO emissions is estimated to be just over 1 million grams. As with energy use, the greatest proportion of CO emissions results from the use of automobiles—in this case about 97%. On a per passenger basis this represents approximately 105 grams per person, compared with approximately 4 grams per person for transit users. As mentioned above for energy consumption, the amount of emissions per person for transit on a daily basis is likely to be higher.

Review

This illustrative design project has shown how the technological characteristics of the transportation system may be designed to respond to future demographic and land-use conditions in an urban area, and how the major impacts may be estimated. Service policies were defined in terms of fare levels and travel time components for transit users, and as a level of service, and hence a travel speed, for automobile users.

As indicated at various points in the analysis, certain approximations were made in order to continue the process to completion, i.e., to obtain estimates of the impacts likely to result if the project were implemented. Clearly, if the impacts were found to be near the threshold of acceptable levels it could be appropriate to refine the analyses in order to be more certain of the level of impacts.

Projects for Solution

This subsection provides eight different sets of policy variables that might be proposed to investigate future air pollution and energy use levels attainable within a transportation corridor. Completion of each project will illustrate the likely ranges of the impacts to be expected. The extent of the work involved in each alternative is such that one person or a team of two should be able to handle the work and conduct some additional exploratory analyses if desired.

Scope of Work

The work to be done for each of the projects is as follows:

1. The corridor area of interest is that shown earlier in Figures 4-1 and 4-3, including the topographic features, traffic zone boundaries, existing transportation system, and demographic data. These data also are the same for each of the eight alternatives. The populations by zone, income, and trip generation data are also the same; these are shown in Figure 4-8.

2. The policy variables for each of the eight alternatives are shown in Table 4-19. The policy variables shown indicate a relatively auto-intensive policy for Alternative 1 through to a "transit-intensive" policy with Alternative 4. The same policy alternatives but with overall lower automobile level of service (LOS) are then typified by Alternatives 5 through 8.

3. The trip generation characteristics and highway and transit networks are the same as those presented in the earlier example, and travel times on each link, travel time factors, and related features are also the same.

The ways in which the policy variables change as the policy favors greater use of transit are as follows:

1. The waiting times for transit users are decreased.
2. The out-of-pocket fares for transit users are decreased.
3. The automobile occupancy is increased from 1.2 to 1.3.
4. The automobile LOS for Alternatives 5 through 8 is decreased from C to E.

Conduct of the Project

It is suggested that the order of conducting the work, the calculations format, and the presentation of interim results of the different demand and design stages be the same as in the project outlined earlier in this chapter. Use of the same format, as is typical of the requirements in a design and planning office, allows the results of each project to be monitored and compared with the others for consistency. This enables errors or inconsistencies in one project to be identified by simple comparison with other project results. A summary sheet to record the lev-

TRIP GENERATION

	Trip Production		Trip Attraction	
Zone	# DUs	Avg Income per DU	# Employees	Trips per Employee
1	50	32	8	16
2	30	24	8	12
3	20	16	9	3
4	5	16	30	12

Note: DUs, Incomes, and Employees are in 1,000s

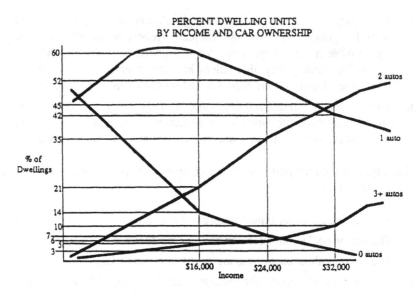

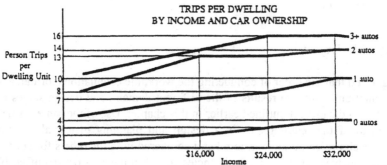

Figure 4-8 Population by zone, income, and trip generation

Table 4-19 Policy Variables for the Alternatives

ALTERNATIVE 1

Zones	Mode	Service Time, S [min]	Line-Haul Time, L [min]	Out of Pocket Costs, C [cents]	Auto Occ.	Auto LOS [vph]
1-4:						
	Auto	5	30	600	1.2	C (1200)
	Transit	25	50	100		
2-4:						
	Auto	5	28	580	1.2	C (1200)
	Transit	20	42	85		
3-4:						
	Auto	5	12	550	1.2	C (1200)
	Transit	20	22	75		

ALTERNATIVE 2

Zones	Mode	Service Time, S [min]	Line-Haul Time, L [min]	Out of Pocket Costs, C [cents]	Auto Occ.	Auto LOS [vph]
1-4:						
	Auto	5	30	600	1.25	C (1200)
	Transit	20	50	80		
2-4:						
	Auto	5	28	580	1.25	C (1200)
	Transit	15	42	70		
3-4:						
	Auto	5	12	550	1.25	C (1200)
	Transit	15	22	60		

ALTERNATIVE 3

Zones	Mode	Service Time, S [min]	Line-Haul Time, L [min]	Out of Pocket Costs, C [cents]	Auto Occ.	Auto LOS [vph]
1-4:						
	Auto	5	30	600	1.3	C (1200)
	Transit	15	50	75		
2-4:						
	Auto	5	28	580	1.3	C (1200)
	Transit	10	42	60		
3-4:						
	Auto	5	12	550	1.3	C (1200)
	Transit	10	22	50		

ALTERNATIVE 4

Zones	Mode	Service Time, S [min]	Line-Haul Time, L [min]	Out of Pocket Costs, C [cents]	Auto Occ.	Auto LOS [vph]
1-4:						
	Auto	5	30	700	1.3	C (1200)
	Transit	10	50	75		
2-4:						
	Auto	5	28	700	1.3	C (1200)
	Transit	10	42	60		
3-4:						
	Auto	5	12	650	1.3	C (1200)
	Transit	10	22	50		

ALTERNATIVE 5

Zones	Mode	Service Time, S [min]	Line-Haul Time, L [min]	Out of Pocket Costs, C [cents]	Auto Occ.	Auto LOS [vph]
1-4:						
	Auto	5	35	600	1.2	E (2200)
	Transit	25	50	100		
2-4:						
	Auto	5	33	580	1.2	E (2200)
	Transit	20	42	85		
3-4:						
	Auto	5	14	550	1.2	E (2200)
	Transit	20	22	75		

ALTERNATIVE 6

Zones	Mode	Service Time, S [min]	Line-Haul Time, L [min]	Out of Pocket Costs, C [cents]	Auto Occ.	Auto LOS [vph]
1-4:						
	Auto	5	35	600	1.25	E (2200)
	Transit	20	50	80		
2-4:						
	Auto	5	33	580	1.25	E (2200)
	Transit	15	42	70		
3-4:						
	Auto	5	14	550	1.25	E (2200)
	Transit	15	22	60		

ALTERNATIVE 7

Zones	Mode	Service Time, S [min]	Line-Haul Time, L [min]	Out of Pocket Costs, C [cents]	Auto Occ.	Auto LOS [vph]
1-4:						
	Auto	5	35	600	1.3	E (2200)
	Transit	15	50	75		
2-4:						
	Auto	5	33	580	1.3	E (2200)
	Transit	10	42	60		
3-4:						
	Auto	5	14	550	1.3	E (2200)
	Transit	10	22	50		

ALTERNATIVE 8

Zones	Mode	Service Time, S [min]	Line-Haul Time, L [min]	Out of Pocket Costs, C [cents]	Auto Occ.	Auto LOS [vph]
1-4:						
	Auto	5	35	700	1.3	E (2200)
	Transit	10	50	75		
2-4:						
	Auto	5	33	700	1.3	E (2200)
	Transit	10	42	60		
3-4:						
	Auto	5	14	650	1.3	E (2200)
	Transit	10	22	50		

Table 4-20 Summary Sheet for Impacts of Each Alternative

Alt. #	Capital Cost [Million $]	Energy [1,000 MJ]	CO emmissions [1,000 grams]	Comments
1				
1				
2				
2				
3				
4				
5				
6				
6				
7				
8				

Comments (Extend as necessary)

_____

_____

_____

els of impacts of each of the eight projects is shown in Table 4-20. This enables the impacts associated with each alternative to be compared with each of the others. It is the foundation of the technical findings used to assist in deciding which projects are acceptable, and for guiding further investigations or refinements of the designs.

References

1. Schoon, John G., "Introduction to Transportation Corridor Planning and Design: A Project Workbook," Course notes, 1986 through 1994, Department of Civil Engineering, Northeastern University, Boston, MA.
2. Transportation Research Board, National Academy of Sciences, *Highway Capacity Manual,* Special Report 209, third edition, Washington, D.C., 1992.

PART 3

Short-Range Transportation Systems Planning

5

Transportation Demand Management and HOV Projects

As described in Chapter 1, Transportation Demand Management (TDM) emphasizes the reduction of demand to require fewer roadway facilities at peak hours. Transportation Systems Management (TSM) focuses on the implementation of relatively low cost, quickly established actions to obtain higher efficiency in moving people as opposed to vehicles. Some overlap in the meaning of these terms makes it advisable to specify the context when using them. The term TSM has been used previously to describe the subject of this chapter, and TDM is applied in most current programs.

This chapter describes the design and planning aspects of estimating the impacts of high occupancy vehicle (HOV) facilities—a recognized method of increasing transportation efficiency and reducing environmental impacts in certain settings. HOV facilities planning and design requires careful investigation of the physical, operating, and demand aspects of transportation modeling described in this chapter and in the associated projects of Chapter 6. HOV planning and design may be conducted separately from, or in conjunction with or following, the long-range planning described in Chapters 2, 3, and 4.

HOV Facilities—Physical and Operating Characteristics

Introduction and Background

In keeping with the emphasis on the major transportation corridor planning, design, and impacts described in the preceding chapters, some of the many freeway-related actions offering significant reduction of congestion and related travel and environmental benefits are listed in Table 5-1. The six categories operating or being implemented in North America and other locations include incident management, highway surveillance and control, motorist information systems, ramp metering, added lanes, and HOV facilities (21). In recognition of the

Table 5-1 Summary of Methods for Alleviating Freeway Traffic Congestion

Congestion reduction tool	Impact	Cost	Implementation
Freeway incident management systems	Could reduce congestion on about 30 % of an urban freeway system; could reduce incident duration by an average of 10 minutes; Benefit/cost of 4:1	$1 million to design and construct; $100,000 maintenance	Long timeframe to implement; requires multiagency approach
Freeway and arterial surveillance and control	Similar to freeway incident management systems, only over wider geographic area	Expensive; few systems in existence today	Multiagency effort required; public education needed
Motorist information systems	Significant reductions in delay on specific facilities	Can be designed for low cost	Long timeframe required; outreach needed to local officials and media.
Ramp metering	Highway speeds increase by 24%; volumes increase from 12 to 40%; 20 to 58% reduction in accidents	Depending on type of system, can be low to moderate cost	Long timeframe; need detailed planning effort to avoid local area problems
Add lanes without widening	Significant increases in capacity possible; Benefit/cost of 7:1	About $1.3 million per mile for design and construction; $12,000 per year for maintenance	Requires joint effort with enforcement agencies; need public education effort
HOV lanes	Potentially significant increases in person-moving capacity; reduce vehicle miles traveled by 5%, and travel times by 6%; Benefit/cost of 6:1	Varies by type; taking an existing lane can be low cost; providing new lanes may cost up to $5 million per mile	Extensive planning required; multi-agency cooperation; need public education and marketing campaign

Source: Ref. (21)

significant reductions in environmental impacts, more efficient facility operations, and lower cost of construction per passenger mile traveled, several states
have mandated the implementation of HOV facilities in transportation corridors
where freeways experience excessive congestion.

Within the context of major urban transportation corridor planning and design,
we will focus on the provision of HOV facilities on freeways—in many respects
an extension of the long-range planning and design described in Chapters 2
through 4. Possible ways of alleviating freeway congestion and some of the
advantages of effectively implemented HOV facilities are described in Figure 5-1
and Table 5-2, respectively. Characteristics of selected HOV and general purpose
(GP) lane configurations are shown in Figure 5-2.

The major objectives of HOV facilities listed by the American Association of
State Highway and Transportation Officials (AASHTO) (1) are to:

1. Maximize person-moving capacity of roadway facilities by providing
    improved operating level of service for HOVs, both public and private

2. Conserve fuel and minimize consumption of other resources needed for
    transportation

3. Improve air quality

4. Increase overall accessibility while reducing vehicular congestion.

Physical and operational planning and design of HOV facilities is assisted by a
number of guidelines prepared by the U.S. Department of Transportation
(USDOT) Federal Highway Administration (FHWA) (2) and others (3–7). However, the methods and techniques of estimating the relative demand for users of
the HOV and associated low occupancy vehicle (LOV) facilities and, hence,
determining their likely economic feasibility and environmental impacts are currently (1996) being developed, and several methods are in use. They are discussed later in this chapter.

Preliminary Guidelines for Implementing HOV Facilities

The decision to proceed with the demand analysis for HOV facilities should only
be undertaken if certain conditions are likely to apply. These conditions include
the presence of congestion prior to the HOV lanes' implementation; significant
travel time savings likely to accrue to users of the HOV lanes (at least 5 minutes
and preferably at least 8 minutes, or at least 1 minute per 1.5 km over mixed-flow
traffic); a vehicle volume of at least 400 vehicles per hour using the HOV lane;
safe operations (particularly for vehicles entering and exiting the HOV lane), and
cost effectiveness.

Moreover, experience has shown that public acceptance has been hard to obtain
if the general purpose (GP) lanes experience greater congestion than before

Busway facility, separate right-of-way

Barrier-separated facility, freeway right-of-way

Buffer-separated facility, freeway right-of-way

Contraflow facility

Nonseparated facility

Queue bypass facility

Figure 5-1   Physical types of HOV facilities   *Source*: Ref. (7)

Table 5-2   Advantages of HOV Projects

- **Protection of Future Person-Moving Capacity:** With HOV lanes, the user requirements can be adjusted as needed to achieve desired service levels. For example, if the HOV lanes are opened to two or more (2+) persons per vehicle and use increases to such a degree that the level of service deteriorates, occupancy requirements can be increased to three or more (3+) persons per vehicle. This flexibility can ensure that the future person-moving capacity of the HOV lanes can be preserved.
- **Implementation Time:** HOV facilities frequently represent the fastest approach for getting some form of fixed transit guideway into operation.
- **Implementation Cost:** Although actual implementation costs are site specific, HOV lanes often represent the least costly fixed-guideway transit facility.
- **Incremental Implementation and Operation:** HOV lanes are amenable to incremental development and can be phased over time, consistent with need and available funding.
- **Cost Effectiveness:** Evaluation of HOV lanes on congested freeways has shown that the benefit/cost ratios for such projects are frequently in excess of six (6).
- **Multiagency Funding:** HOV facilities are often eligible for local, state, and federal funding from both highway and transit agencies.
- **Multiple User Groups:** In addition to transit bus vehicles, vanpools and carpools can use the available capacity in the HOV lane, thereby increasing the total person-moving potential.
- **Schedule Reliability:** Transit service can be made more reliable, permitting more efficient scheduling services. In some circumstances, transit vehicle productivity can be improved as well.
- **Operating Speed:** Transit service on priority lanes is often express and nonstop. As a result, the average travel speeds are extremely high.
- **Operating Cost:** Carpools, vanpools, and transit vehicles all realize significant cost savings by traveling at higher speeds in HOV lane than in the mixed-flow freeway lanes during periods of congestion.
- **Flexibility:** HOVs can use the existing street system for collection and distribution.
- **Reinforcement of Activity Center Land Uses:** A major portion of future traffic will be attributable to these major activity centers, and HOV lanes can enhance service to and from them, thus reinforcing existing and projected development.
- **Increased Reliance on Private-Sector Modes:** Preferential treatment promotes the use of a greater number of transportation modes, including private carriers, vanpools, and carpools.
- **Time-Adjustable Operation:** On some HOV facilities, particularly nonseparated-flow and contraflow lanes, the lane used by priority vehicles during peak periods can be used for other beneficial purposes (e.g., extra mixed-flow lane) during the nonpeak period.

Source: Ref. (7)

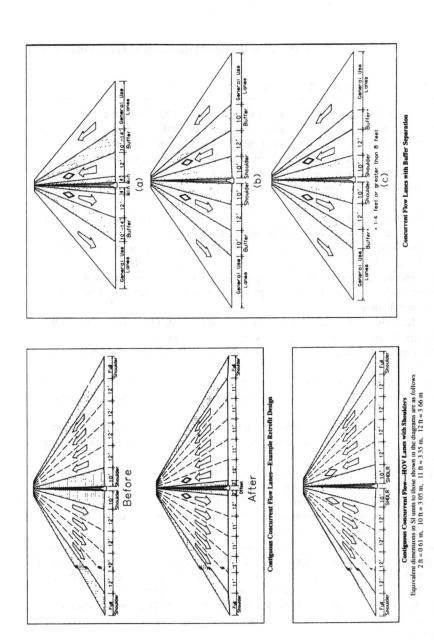

Figure 5-2 Selected types of freeway HOV lane configurations   *Source*: Ref. (1)

implementation of the freeway HOV lane. This would occur under the "take-a-lane" case where one lane is converted to HOV use, thus reducing the number of GP lanes. This has resulted in the general guideline of not reducing the number of GP lanes and, instead, adding an HOV lane, either contiguous or contraflow, on a separate way. However, recent investigations (20, 23) indicate that implementing the add-a-lane method may result in a greater number of trips and vehicle miles traveled (VMT) with resulting total increases in air pollution (19) and energy impacts, although there could be a net overall reduction in person delay, depending on the nature and extent of the congestion. The increase in trips would be due to improved accessibility, thus inducing additional traffic. The investigations (20) also found that the public may more readily accept the implementation of take-a-lane HOV projects if they are properly informed of the likely outcome beforehand.

HOV Analysis Procedures

Two general approaches to the demand analysis process mentioned by the AASHTO are (1) modified regional mode-choice models and, (2) the microcomputer-based and/or manual freestanding procedures, usually applied to transportation corridors. Use of each of these demand models assumes that initial physical and operational features of the proposed HOV facilities appear to be broadly acceptable. The regional demand estimation models typically adopt the techniques of the traditional transportation demand modeling system—essentially similar to the analysis methods described in Chapters 2 through 4. They would include trip tables, origin and destination data by purpose, trip length, and the results of modal split and traffic assignment analysis throughout the region. The latter two stages can incorporate vehicle occupancy by bus, car/vanpools and LOVs into modal split and traffic assignment procedures that are sensitive to travel speed differentials resulting from the HOV facilities on any given travel network link, and to associated user costs.

The freestanding corridor level HOV models vary considerably in their complexity and analysis techniques. These approaches are often used in the sketch planning or screening levels of the analysis but may be developed in greater detail to justify implementation of a particular project. Nevertheless, it is desirable that the freestanding models be verified and the results refined by using an areawide or regional model before final planning and design. These models may use data obtained extensively in the relevant travel corridor, such as lane volumes, vehicle occupancies, recorded travel speeds, and volume/density characteristics. They may also use traffic assignment volumes based upon the modal split and trip tables, as well as data typical of quick response methods such as those described in National Cooperative Highway Research Reports 186 and 187 (8, 9). Several publications provide comparisons between the various planning methods related to HOV facilities, (10–13), and to evaluating the effectiveness of HOV facilities (14, 15), (16).

The approach to exploring the impacts of proposed HOV projects described here is a freestanding method for screening and preliminary investigation of HOV projects in a radial or circumferential corridor of a metropolitan area. The approach addresses demand estimation to the extent needed to identify the major reasons for the project's inclusion in a list of candidates, or to assist in estimating approximate regionwide impacts. Although the description refers specifically to concurrent freeway HOV facilities, it is also relevant, with modifications, to other freeway HOV situations.

Major Analysis Elements

Between the initial concept of providing HOV facilities on a particular freeway route and the eventual HOV implementation, the corridor and the route should be evaluated for physical and operational characteristics using the following steps:

1. Identify the areas of congestion in terms of traffic speed, volume, and density and associated impacts such as air pollution and mobility, and other, often subjective or community-related concerns such as potential diversion of traffic through residential neighborhoods.

2. Identify the likely breakdown between potential HOVs such as buses, carpools, vanpools, etc., to determine if adequate HOV users are likely to be available.

3. Check for physical and operational features such as access and egress locations for the HOV vehicles, possible need for weaving across freeway lanes, levels of enforcement, sufficient lanes and physical space for additional lanes, space for special on and off ramps, and related safety and operational features.

If the above conditions appear satisfactory, the more promising freeway segments for implementation are subject to a quantitative screening analysis for the anticipated demand and impacts (the major emphasis of this chapter). This analysis examines the likely lane volumes and demands and the levels of service (LOS) indicated by travel speeds and other traffic flow features. The final stage is a detailed analysis to ensure that the proposed changes are consistent with regional or corridor-wide plans, air pollution mandates, and other objectives. This latter aspect of the planning and design process is beyond the scope of our discussion.

Some of the major physical and operational concerns described above are shown conceptually in Figure 5-3(a), and a flowchart showing major elements of implementing HOV facilities is shown in Figure 5-3(b). Highlighted in the latter

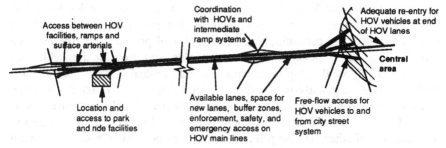

(a) SELECTED PHYSICAL AND OPERATIONAL REQUIREMENTS FOR HOV FACILITIES - DIAGRAMMATIC

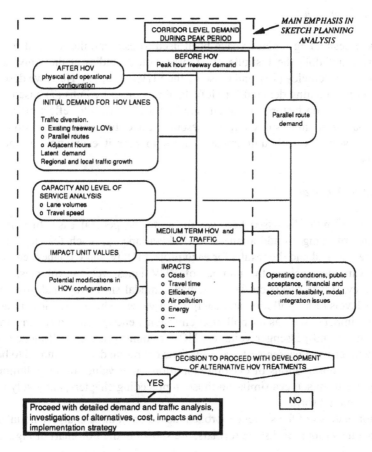

(b) OUTLINE OF INITIAL DEMAND ANALYSIS PROCESS FOR HOV FACILITIES

Figure 5-3   Initial considerations in determining HOV facilities feasibility

are the main areas of focus in this and the next section: corridor level demand, HOV demand estimation, traffic flow analysis, and estimation of impacts. These elements combine to determine whether a proposed project meets the required planning guidelines, including impact thresholds, and provides a checklist for "fatal flaws" before developing alternatives and eventual implementation. Normally, the screening analysis would be conducted for the peak hour of freeway demand. This observed demand during the peak period reflects the peak *hour* demand (the usual basic analysis period), which is often affected by demand in the peak *period* other than the peak hour, demands on parallel routes, and latent demand.

Analysis Approach

Several freestanding methods have been used to estimate the demand for HOV facilities, including the Institute of Transportation Studies at the University of California at Berkeley (13), and those by the FHWA (10, 17). Here, we describe a method of estimating demand for HOV facilities given by Fuhs (7) based upon a range of diversion factors, and augmented by Schoon (22) to reflect modal shift, lane capacity, running speeds, total passenger travel time, and integration of the demand estimates with unit impact values to permit estimation of the total impacts.

Analysis Methods

In the example discussed here we illustrate the general case. In Chapter 6 examples of using an add-a-lane and a take-a-lane approach for an A.M. peak hour are outlined and the results for each project (HOV and GP lanes) compared.

The variables include the number of HOV lanes, vehicle volumes, vehicle occupancy, traffic diversion, latent demand (modal shift), regional growth, and various levels of air pollution and energy unit values. These variables are used to estimate impacts such as air pollution emissions, energy use, HOV lane and GP lane vehicle and passenger volumes, HOV running time, and efficiency of the facility in terms of passenger travel time. The operating data is assumed to be collected from observations of existing conditions, unit values of air pollution and energy use from sources similar to those in preceding chapters, and likely ranges of future modal shift.

Major features of the subject approach, based upon three analysis modules that incorporate various worksheet formats, are shown in the flowchart of Figure 5-4.

Module 1—Existing (Before HOV) Conditions

The format of the worksheet is shown in Table 5-3. The input values are entered into the boxes shown in the worksheet and the calculations are performed in a

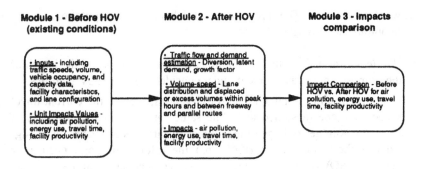

Module 1 - Before HOV (existing conditions)	Module 2 - After HOV	Module 3 - Impacts comparison
• **Inputs** - including traffic speeds, volume, vehicle occupancy, and capacity data, facility characteristics, and lane configuration • **Unit Impacts Values** - including air pollution, energy use, travel time, facility productivity	• **Traffic flow and demand estimation** - Diversion, latent demand, growth factor • **Volume-speed** - Lane distribution and displaced or excess volumes within peak hours and between freeway and parallel routes • **Impacts** - air pollution, energy use, travel time, facility productivity	**Impact Comparison** - Before HOV vs. After HOV for air pollution, energy use, travel time, facility productivity

Figure 5-4 HOV impacts analysis process

Table 5-3 Before HOV Impacts Estimation Worksheet

All volumes are peak hour volumes unless otherwise noted.

MODULE 1: BEFORE HOV, INPUTS AND IMPACTS
All volumes are per hour unless stated otherwise

Item	Info. Source, or Calculation	HOVbus	HOVauto 2+	SOV	Total
Inputs: All 4 lanes (values per lane):					
Current (before HOV) travel speed, km/h	Surveys				
Vehicle volume by mode	Surveys, check consistency with speed				
Vehicle mode split, %	Mode % of total volume				
Check total veh vol, pcphpl	Lookup la vol pcphpl × 4 . . . (4 × 1400)				
Lane volume, pcphpl	HCM value divided by number of lanes				
Ave. Vehicle Occupancy	Surveys				
Number of passengers	Vehicles × ave. occupancy				
Pass. mode split, %	Mode % passengers of total				
HOV section length, km.	Measured				
Impacts:					
HOV veh running time, hours	HOV segment length/speed				
Emissions (CO), gm/veh km	Lookup				
Emiss/mode	Section length × vol. × emiss/veh km				
Energy use, MJ/km	Lookup				
Energy use/mode, MJ	Length × vol. × MJ/veh km				
Passenger travel time (hours)	Length × passengers/speed				
Pass. modal split, % HOV users	All HOV pass/Total passengers, %				

downward progression. The input data, highlights of the computational method, and the outputs are described below.

- *Physical data* include the total number of freeway lanes and the length of the proposed highway segment under consideration for HOV facilities.
- *Operational data,* obtained from observations, include the current travel speed by mode (assumed to be the same for all vehicles regardless of lane), the volumes of buses, the volumes of carpools and vanpools (designated in the tabulations as HOVautos), the volumes of single occupancy vehicles (SOVs), and the average vehicle occupancy of each mode. Because the number of buses that will divert to the HOV lane is usually much less than the HOV lane's capacity, no attempt is made here to convert the bus volume to equivalent passenger cars. A visual check is made to ensure that the observed vehicle speeds are approximately the same as the speed-volume-capacity relationships described in (18), shown graphically in Figure 5-5, and tabulated in the Appendices. Note that for analysis purposes the relationship between speed and flow is based on an assumed curve for the interrupted flow conditions of LOS F. The curve is based on data shown in (18), Chapters 2 and 3, and other sources. The lane volume value from (18) is used from this point in the ensuing analyses to ensure that the traffic flow relationships between speed and volume are consistent, when the After HOV conditions are analyzed.
- *Impact data.* In this case the unit values for air pollution emissions and energy use are entered into the worksheet. The values are based upon the general relationships described in Chapter 3, and tabulated in the Appendices.
- *Analysis and outputs.* Estimating the impacts is an arithmetic process, shown in the worksheet, resulting in estimates of the HOV vehicles' running time, total emissions, energy use, and passenger travel time, and of the segment of freeway under Before HOV (existing) conditions. These impacts provide the basis for comparison with the After HOV impacts described below.

Module 2—After HOV

As shown in Table 5-4, the worksheet process shown in this module includes all of the assumptions about the future demand characteristics, and calculates the resulting impacts. Module 2 is broken down into three parts: demand estimation, traffic flow analysis, and estimation of impacts, each of which is described below.

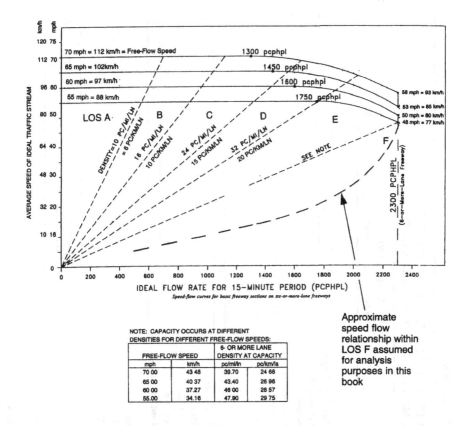

Figure 5-5 Volume-capacity relationships   *Source*: Ref. (18), Figure II. 3–1, modified
to also show speeds in SI units and assumed speed-flow relationship within LOS F.

Demand estimation

The process described here is based upon a conservative approach that incor-
porates a method described in (7) and also on published elasticity values relating
travel time to modal shift. The description applies to the case of 2+ HOVs
(meaning that vehicles with two or more occupants are classified as HOVs). The
procedure for estimating 3+ HOVs is similar except that different input values
may be applicable. However, little definitive historical information is available at
this time (1996) regarding 3+ operations, and they are not discussed further here.
Demand estimates for a prospective HOV facility must address three sources of
vehicles and passengers using it:

Table 5-4. After HOV Demand, Traffic Flow, and Impacts Worksheet

MODULE 2: AFTER HOV TRAFFIC FLOW, DEMAND, AND IMPACTS-ADD-A-LANE

Item	Info. Source, or Calculation	LANE 1 (HOV) HOVbus	HOVauto	LANES 2-5 (GP) Bus	HOVauto	SOVs	TOTAL
Demand Estimate:							
Diversion factor	% of eligible bus and HOVauto to HOV lane		☐				
Pass. vols. following diversion	Factor the "before HOV" pass volumes						
Modal shift from SOVs	% of SOVs to bus and HOVautos					☐	
Pass. demand vols. from mode shift	Add shift to HOVs and deduct from SOVs						
Pass. volume by mode, lane	Add mode shift pass. to previous pass.						
Vehicle volume by mode	Pass. volume/veh. occupancy						
Traffic Flow Analysis:							
Demand vols. (per lane basis)	Bus+HOA in HOV lane, and ea. GP lane	☐		☐	☐	☐	
Gen Purpose la. forced flow capacity	HCM value under 'before' conditions						
GP lanes, excess capacity, per lane	Demand vol. by lane - forced flow capacity						
GP lanes excess capacity, total	Lane difference × number of lanes						
Actual vols. by mode, per lane	GP lanes: split cap. vol. between modes	☐		☐	☐	☐	
After HOV speeds, km/h	Lookup (manual) HCM 94 and lane vol						
Impacts by Lanes:							
HOV veh running time, hours	HOV segment length/speed						
Vehicle volumes	Vehicle volumes by category						
Passenger volumes	Mode vol × veh. occupancies						
Emissions (CO), gm/veh km	Lookup	☐		☐	☐	☐	
Emiss/mode	Length × vol. × emiss/veh km						
Energy use, MJ/km	Lookup	☐		☐	☐	☐	
Energy use/mode, MJ	Length × vol. × MJ/veh km						
Passenger travel time (hours)	Length × passengers/speed						
Pass. modal split, % HOV users	All HOV pass/Total passengers, %						

All volumes are peak hour volumes unless otherwise noted.

1. Corridor growth
2. Existing HOVs diverted from the mainline freeway and from parallel routes
3. Modal shift (latent demand).

Corridor traffic growth is subject to local factors and is best developed based on locally adopted growth projections or, ideally, growth projections for specific zonal activity similar to that illustrated in Chapter 4. If no growth projections have been developed, an appropriate option in most locations is to apply historic growth trends. This is generally suitable because the implementation period for HOV projects can be quite short, and a determination of concept viability is not likely to be altered by the inclusion of this factor.

Diversion refers to the potential for existing eligible HOV lane users to divert into the HOV facility and use it. Most diversion occurs from the freeway mainline or from service roads adjacent to the HOV lane. For purposes of this methodology, this is called *primary diversion*. Diversion that is likely to occur from adjacent, often parallel, streets and highways is termed *secondary diversion*. Based on (7), primary diversion may range from 70 to 90% of HOV vehicles in the Before HOV case. Secondary diversion may range between 1.2 and 1.6 times the primary diversion. Separate factors are applied for each.

Modal shift, or latent demand, is less corridor and project specific and can be best assessed from the experiences of other HOV projects. The generalized criteria presented herein are suitable as long as a travel time savings threshold of 5 to 8 minutes is met over the study limits. If not, a modal shift cannot be expected to occur that would represent the estimation factors provided. For the purposes of an initial study, *modal shift* is defined as a shift to HOV vehicle use by commuters currently driving alone but attracted by the potential HOV lane's time savings. These shifts can occur from lower occupancy vehicles or from vehicles not currently making the trip. For purposes of this book, modal shift will be based upon the elasticity values represented graphically in Figure 5-6. Here, the changes in the independent attributes (in this case the travel time difference between using the HOV lane versus the GP lane) are related to the corresponding modal shift.

Few HOV projects have operated long enough to fully document factors that influence use, and even fewer have undertaken extensive before-and-after data collection. Because only a limited data base exists from which criteria can be drawn, caution should be used in applying modal shift factors.

The demand estimation ends with the redistribution of the number of vehicles and passengers between the mode categories, i.e., buses and HOVautos in the HOV lane, and the remaining HOVbuses, HOVautos, and SOVs in the GP lanes. The buses and HOVautos in the GP lanes are those remaining after the diversion.

Another approach that estimates the total After HOV volumes in the peak hour could be to adopt a method developed by Parody (10) resulting from investigating

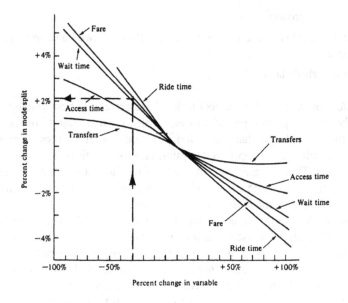

Example of Using the Chart

Problem
Assuming that the fare is reduced by 30% for a specific transit trip, what is the estimated change in modal split (i.e., the modal shift)?

Solution
Follow the dashed line from the percent change in variable (in this case the fare) to the curve for "fare" and read the change in modal split as +2%, i.e., because the fare is reduced by 30%, the increase in passengers is estimated to be 2%.

Figure 5-6   Sensitivity of change of mode to changes in variables

vehicle modal shifts for new HOV facilities in the United States. This reference provides examples of the use of this method, based upon regression analyses incorporating lane use, travel time, cost, and traffic flow.

Traffic flow analysis

The effects of the redistribution of vehicle volumes in the new HOV lane and GP lanes will include the following:

1. The HOV lane will typically operate at a volume well below capacity, and, providing that no downstream congestion exists, a corresponding increase in travel speed and a reduced travel time will result.

2. The traffic in the GP lanes will experience flow conditions that vary fundamentally depending on whether an add-a-lane or a take-a-lane HOV facility is implemented. For *both* conditions, two basic assumptions are implied: First, the congested flow during the peak hour extends through the hours immediately preceding and following the peak hour. Second, the lack of downstream capacity (a bottleneck) causes the forced flow conditions typical of LOS F in the GP lanes. Both of these assumptions are somewhat idealized, because only limited documentation is available on the lane volume redistribution of GP lanes under forced flow conditions associated with HOV facilities. Some of the aspects of these characteristics are addressed by Schoon (22). However, under the foregoing assumptions the following traffic conditions are likely to occur in the GP lanes:

    (a) *Add-a-lane case.* The shift of HOV vehicles to the added HOV lane will allow vehicles from the hours immediately preceding and following the peak, and possibly from traffic diverted from parallel routes, to travel in the peak hour. However, because of the assumed continuing downstream bottleneck, the volume and associated speed in *each* GP lane will remain essentially unchanged in the peak hour. This assumption is conservative in that, preferably, improvements to downstream bottlenecks would be implemented in conjunction with the HOV program so that the total lane flows and speeds will increase. However, in practice, this is often difficult to achieve. In the long run, increase in total capacity of both HOV and GP lanes will attract a greater volume of trips because of the improved accessibility, and the volumes immediately preceding and following the peak hour will eventually revert to their original volumes, resulting in a greater total volume of vehicles during the peak period.

    (b) *Take-a-lane case.* In this case, the conversion of at least one lane to an HOV lane, with no increase in the total number of lanes, results in a reduction in the number of GP lanes. Again, because of the assumed continuing downstream bottleneck, the volume and associated speed in each GP lane will essentially be unchanged in the peak hour. These conditions will result in a net *decrease* in the total capacity of the GP lanes during the peak hour, and the demand will therefore tend to be *displaced from* the peak hour to the hours immediately preceding and following it, and to parallel routes if capacity is available there—in contrast to the add-a-lane case where traffic is *absorbed by* the peak hour. The long-run implication is that the accessibility offered by the corridor will experience a net decrease during the peak hour and period (in terms of vehicle capacity, but not necessarily person capacity if increasing number of passengers shift to the HOV facilities), as opposed to an increase, as in the add-a-lane case.

3. At this point the eligibility for HOV vehicles (2+ versus 3+ occupants per vehicle) may have to be reassessed. *It should be noted that, in practice, no actual peak hour decrease in volumes on the parallel routes may be discernible, because of an excess of demand.*

The final step in the traffic flow analysis is to insert the lane speeds corresponding to the lane volumes, based on the appropriate values in (18) and the demand and traffic flow volumes described above. The speeds provide a direct input to the estimation of impacts in Module 3, described next.

Estimation of impacts

The impacts of the After HOV condition are estimated by extending the new traffic volumes for each mode for the relevant speeds by the same unit impact values used in the Before HOV condition. Also, an estimate of the new modal split is made, expressed as the percentage of passengers using HOV modes of the total volume of passengers.

Module 3—Comparison of Impacts

The differences between the impacts of Module 1 and those of Module 2 are calculated, as shown in Table 5-5. They provide the estimate of the impacts expected if the HOV facilities were implemented, and are the major output of the initial planning process. The impacts are presented both as total values and on a per passenger basis where applicable. The total values are important because they provide an indication of the total impacts in the peak hour. The per passenger basis is useful because it indicates the likely effectiveness of the proposals related to total anticipated demand in the peak period, as well as the demand in only the peak hour. Also shown is the change in modal split, in terms of the change in percentage of passengers using the HOV modes. All of these values are normally developed for the peak hour. Thus, they do not include the benefits or drawbacks associated with reducing or increasing volumes in the adjacent hours. The impacts associated from adding volumes to succeeding hours (as in the take-a-lane case) may more than offset the benefits achieved during the peak hour.

Essentially, the comparison of impacts will indicate whether the proposed project appears to address sufficiently the issues of concern. The alleviation of con-

Table 5-5 Comparison of Impacts

	Module 3: Differences between Before HOV and After HOV Impacts			
	Total change		Change per passenger	
Impact item	Amount	Percentage	Amount	Percentage
HOV vehicle running time, hr.				
Vehicle volume, veh./hr.				
Passenger volume, Pass./hr.				
CO emissions, gm.				
Energy use, MJ				
Passenger travel time, total hours				
Pass. modal split, all HOV users (incl GP las.)				

gestion in order to improve mobility and access is important. But often this must be accompanied by stabilization or, preferably, reductions in adverse environmental impacts. The method shown has described a framework within which the variables may be assembled and their effects explored to provide decision makers with essential information about how to proceed.

Review

The foregoing outline of the analysis method for initial planning of HOV facilities has shown an approximate method of estimating the impacts based upon available inputs, worksheet computational methods, and direct application of unit impact values to the speeds and volumes of the various vehicle modes.

Although the method presented is quite basic in approach, it may be refined in several ways. These include the following:

1. Different physical and operational features can be explored by changing the values and parameters associated with, for example, additional lanes added to the freeway, either concurrent or contraflow, by modifying the format somewhat. Also, the use of 3+ or other car/vanpool occupancies eligible for using the HOV lanes can be explored.

2. Input variables can be modified to reflect values of vehicle occupancy, volume/density/speed relationships, and a variety of scenarios related to diversion, latent demand (modal shift), and growth. Modal shift usually changes over a period of time. Thus, the analyst can experiment with a range of values to see what the overall effects may be. By constructing a

number of relationships of this type the analyst can explore a range of impacts likely to occur from a given set of scenarios associated with the specified variables.

3. If a more detailed analysis of modal shift is available, it can be incorporated into the overall process simply by inserting the appropriate values.

4. Unit values of impacts used for this analysis are basic, approximate values that do not take into account, for example, the effects of emissions and energy use with cold engine temperatures. Also, the effects of stops and starts are not included, and a factor can be incorporated in the analysis to account for these.

5. Values of diverted traffic, latent demand, and traffic growth can be determined for individual cases. Also, other factors such as those representing the extent of compliance with HOV regulations can be included.

It should be noted that, in order to complete a more detailed analysis, the calculations should be conducted for the hours preceding and following the peak hour, all within the peak period, to account for the temporal "transfer" of traffic. Also, consideration should be given to the effects on the parallel route traffic, including public transportation, if these are a significant feature of the travel corridor. Usually, in order to conduct these analyses with reasonable accuracy, a regionwide transportation modeling effort is required, incorporating many of the features presented in earlier chapters. Such an effort is beyond the scope of this book.

Based upon this brief overview of the range of TSM actions, together with the more detailed description of one method of conducting an analysis for a freeway HOV facility, examples of an add-a-lane and a take-a-lane case and a set of problems for solution are provided in the next chapter.

References

1. American Association of State Highway and Transportation Officials, *Guide for the Design of High-Occupancy Vehicle Facilities,* Washington, D.C., 1992.

2. Batz, T. M., *High-Occupancy Vehicle Treatment, Impact and Parameters,* Report Number FHWA/NJ-86-017-7767, Federal Highway Administration, Washington, D.C., 1986.

3. Christiansen, Dennis L., *High-Occupancy Vehicle System Development in the United States, A White Paper,* Texas Transportation Institute, College Station, TX, December 1990.

4. Fuhs, Charles A., NCHRP Synthesis 185 *Preferential Lane Treatment for High-Occupancy Vehicle,* Highway Research Board, National Research Council, Parson Brinckerhoff, Quade and Douglas, Inc., Orange, CA, 1993.

5. Institute of Transportation Engineers, *Guidelines for High-Occupancy Vehicles (HOV) Lanes, A Recommended Practice,* Publication No. RP-017, Washington, D.C.: ITE, 1986.

6. ———, *The Effectiveness of High Occupancy Vehicle Facilities,* Washington, D.C.: ITE, 1986.

7. Fuhs, C. A., *High-Occupancy Vehicle Facilities: A Planning, Design, and Operation Manual,* 189 William Barclay Parsons Fellowship, Parson Brinckerhoff, Inc., New York, NY, 1990.

8. Sosslau, A. B., A. B. Hassam, M. M. Carter, and G. V. Wickstrom, *Travel Estimation Procedures for Quick Response to Urban Policy Issues,* National Cooperative Highway Research Program Report 186, Transportation Research Board, Washington, D.C., 1978.

9. ———, *Quick-Response Urban Travel Estimation Techniques and Transferable Parameters, User's Guide,* National Cooperative Highway Research Program Report 187, Transportation Research Board, Washington, D.C., 1978.

10. Parody, E., *Predicting Travel Volumes for HOV Priority Techniques: A User's Guide,* Report No. FHWA-RD-82-0842, Charles River Associates for the Federal Highway Administration, Boston, MA, 1982.

11. Rothenberg, M. J., and D. R. Sandahl, *Evaluation of Priority Treatments for Highway Occupancy Vehicles,* Report No. FHWA-RD-80-2, Federal Highway Administration, Washington, D.C., 1981.

12. Southworth, F., and F. Westbrook. *High-Occupancy Vehicle Lanes: Some Evidence of Their Recent Performance,* Transportation Research Record 1081, Transportation Research Board, Washington, D.C., 1986:31–39.

13. Scapinakis, Dimitris A., and May, *Demand Estimation, Benefit Assessment and Evaluation of On-Freeway High-Occupancy Vehicle Lanes: Level II, Quick Response Analysis,* Prepared for the California Department of Transportation by the Institute of Transportation Studies, Berkeley, CA, 1989.

14. Turnbull, Katherine F., Robert Stokes, and Russell H. Henk, "Current Practices in Evaluating Freeway High-Occupancy Vehicle Facilities," Texas Transportation Institute, Presented at the Transportation Research Board 70th Annual Meeting, Washington, D.C., January 1991.

15. ———, *Study of Current and Planned High-Occupancy Vehicle Lane Use: Performance and Use,* Report Number ORNL/T3-9847, Oak Ridge National Laboratory, Oak Ridge, TN, 1985.

16. Ulberg, C., *Evaluation of the Cost-Effectiveness of HOV (High Occupancy Vehicle) Lanes,* Report No. WA-RD-121.2, Federal Highway Administration, Washington, D.C., 1988.

17. Gilbert, Keith, *Transportation Systems Management: Handbook of Manual Analysis Techniques for Transit Strategies (Dallas, Fort Worth),* United States Department of Transportation, Urban Mass Transportation Administration Report No. TX-09-0045-81-1, Washington, D.C., 1981.

18. Highway Research Board, *Highway Capacity Manual,* Special Report 209, third edition, National Research Council, Washington, D.C., 1994.

19. Leman, Christopher K., "Does HOV Lane Construction Really Clean the Air?" *Newsline,* Vol. 18, No. 3. Transportation Research Board, Washington, D.C., September, 1992.

20. Jovanis, Paul, *Taking a Look at General Purpose Lanes for HOV Use,* Conference Workshop, Transportation Research Circular No. 409, Transportation Research Board, Washington, D.C., 1992.

21. Institute of Transportation Engineers, *A Toolbox for Alleviating Traffic Congestion,* I.T.E., Washington, D.C., 1989.

22. Schoon, J. G., *"Development of Simplified Screening Method for Investigating HOV Facilities Impacts,"* Paper prepared for submission to Transportation Research Board for Annual Meeting, Washington, D.C., 1997.

23. Bieberitz, John A., "The Effect of HOV Lanes in Reducing VMT—A Theoretical Discussion," Transportation Research Board, *Proceedings of Conference on HOV Facilities,* Los Angeles, CA, June 1994.

24. Voorhees, A. M., and Associates, *Handbook for Transportation System Management Planning—Volume 2, Handbook for Evaluation of Individual Transit-Related TSM Actions,* North Central *Texas.* Council of Governments, 1977.

6

Examples and Projects—HOV Corridor Planning and Impacts Estimation

To illustrate the process and the typical results expected from a proposed high occupancy vehicle facility, this chapter provides examples, followed by a set of projects for solution including variables that may be typically encountered in practice.

High Occupancy Vehicle (HOV) Corridor Examples

These examples illustrate a method of estimating traffic flows and impacts associated with HOV facilities using the freestanding analysis approach described in Chapter 5. The intent is to illustrate a simplified method enabling planners and designers to explore the effects of the key policy variables at the sketch planning level of analysis via manual worksheet calculations or commercial microcomputer spreadsheet programs. The variables include the number of HOV lanes, vehicle occupancy, demand, regional growth, modal shift, and various levels of air pollution and energy unit values. First, an add-a-lane example is shown; then a take-a-lane case using the same Before HOV condition is presented and the results compared.

The examples show how the effects of physical and operational configurations of potential HOV facilities can be investigated to provide relatively simple tabular and graphical representations of the impacts. These summaries can then be used to determine further investigations and alternatives for a potential project.

Physical and Operational Conditions and Assumptions (Add-a-Lane Case)

The major features of the physical and operating conditions of the before and after HOV cases used in the example are summarized in Table 6-1. Also shown is

---

* This chapter is set with a unique style to emphasize the practical nature of the material covered. They are typical project calculations done manually in a design office.

Table 6-1 Existing and Proposed HOV Corridor Conditions (Add-a-Lane Case)

Item	Value or Condition for example[1]		Other possible options, and comments
	Before HOV	After HOV	
Total number of lanes one way	4	5	May vary up to 6 or more lanes
Number of HOV lanes one way	0	1	2 maximum
Number of LOV (GP) lanes one way	4	4	
HOV lane configuration	Not applicable	Concurrent, contiguous	May comprise contraflow, exclusive, permanent, or temporary configurations
Modes and lane apportionment	Bus and car/vanpools, (HOVs) and low occupancy vehicles (LOVs) distributed uniformly throughout the 4 lanes	HOVs in 1 lane, LOV's in 4 GP lanes	Usually, at least 3 LOV lanes are required
Vehicle occupancies			
Bus	40	40	Varies, approximately 20–50
High occupancy automobiles	3	3	Varies, approximately 2–6
Single occupancy autos	1	1	Not applicable
Automobile occupancy qualifying for HOV lane use	Not applicable	2+	3+ or even 4+ may be applicable, depending on relative volumes in HOV and LOV lanes
HOV lane vehicles' speed	Not applicable	90 km/h	Approx. value to assist in approx. estimate of mode shift

GP lanes' travel speeds	24 km/h in all 4 lanes	HOV lane and LOV lanes speeds vary with modal split	Usually, speeds of less than 48 km/h are required to warrant implementation of HOV facilities
Freeway length considered for HOV operations	8 km	8 km	Usually 2 or more miles are necessary for a minimum effective length
Average trip length in travel corridor, including HOV segment	24 km	24 km	Trip lengths typically vary depending on suburban land-use patterns and extent
Approximate average trip time savings for HOV lane users over existing trip time	Not applicable	8 minutes (approx 40% of total trip time)	Assists in indicating initial savings feasibility and in estimating potential modal shift. Assumes HOV speed = 90 km/h
Exit, or downstream traffic conditions	All lanes congested (24 km/h, LOS F)	HOV lane is free flowing. GP lanes remain uncongested with speed = 24 km/h	Unless HOV lane exits to local streets or terminal facilities are free flowing, the full benefit of the HOV lanes cannot be realized,

(1)For traffic flows, vehicle speeds and occupancies, etc., see other data in the information that follows.

an indication of how the values used relate to other possible options. These options are described in the references of Chapter 5.

The project involves a freeway with four general purpose inbound lanes to the central business district (CBD) (**Before HOV**) which will be converted to one HOV lane and four LOV lanes, giving a total of five lanes for the **After HOV** condition. The After HOV configuration will be that of a contiguous, concurrent HOV lane, containing both buses and car/vanpools (HOVautos), with the LOV vehicles operating in the four general purpose (GP) lanes. This is an add-a-lane configuration. Its implications compared to those of a take-a-lane configuration were discussed briefly in Chapter 5. To avoid unnecessary complication, no significant interaction with parallel highway or transit routes is assumed to occur in this illustrative example.

It is considered desirable initially to assume that 2+ persons per vehicle will qualify the vehicle as an HOV. It is assumed that current traffic is flowing at a speed and density well under capacity and within LOS F, or forced flow conditions, throughout the length of freeway route for which consideration is being given (8 km). Also, it is assumed that: (1) the HOV lane's exit from the freeway into the city center will be free flowing, (2) HOV vehicles not destined for the city center will not impede those that are, and (3) no downstream adverse congestion effects will be transferred to the HOV segment. Furthermore, it is assumed that the downstream capacity of each GP lane will not increase during the peak hour being considered, and that the associated travel speeds, volumes, and densities will remain about the same as in the Before HOV condition. During the peak hour, because of HOVs being accommodated in the new HOV lane, some transfer of LOV vehicles and passengers to the GP lanes may occur from the preceding and following hours of the peak period. An example of the lane configuration for this example is shown in Figure 6-1. Also shown is the configuration for the take-a-lane case, presented later. The sections that follow describe the investigation of the two proposals in greater detail.

Objectives

In order to adequately estimate the major impacts of implementing the HOV lane, as well as the associated changes in the other lanes of the freeway, the objectives of this project are summarized as follows:

1. *Travel time savings.* Estimate the travel times of users of the corridor by each of the modes, expressed as total travel time and travel time per passenger.

2. *Facility performance.* Estimate the vehicle running time in the HOV lane.

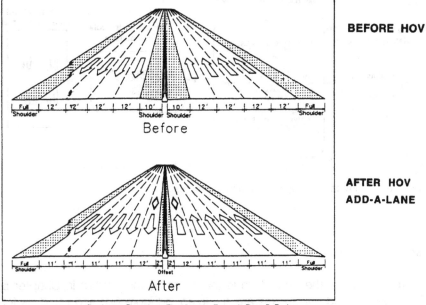

BEFORE HOV

AFTER HOV
ADD-A-LANE

Contiguous Concurrent Flow Lanes—Example Retrofit Design

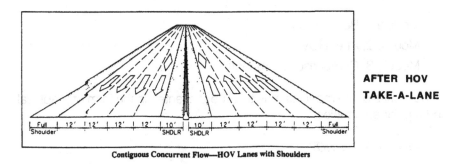

AFTER HOV
TAKE-A-LANE

Contiguous Concurrent Flow—HOV Lanes with Shoulders

Equivalent dimensions in SI units to those shown in the diagrams are as follows:

2 ft = 0.61 m,   10 ft = 3.05 m,   11 ft = 3.35 m,   12 = 3.66 m

Figure 6-1 Lane configuration for proposed HOV facilities   *Source*: Based upon Ref. (1)

3. *Air pollution emissions.* Estimate the amount of CO emitted throughout the corridor.

4. *Energy use.* Estimate total energy use by all modes of transportation in the corridor, measured by fuel consumption, expressed in common units.

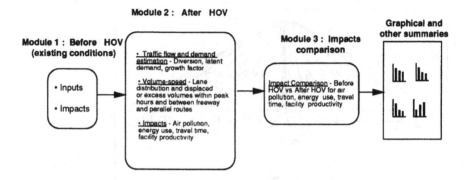

Figure 6-2 Flowchart—Impacts estimation process

Estimation of Impacts

For this example, the procedure is the same as that shown in Chapter 5. The modules and summaries are depicted in Figure 6-2 and listed as follows:

- Module 1, Before HOV
- Module 2, After HOV
- Module 3, Comparison of impacts.

The results of the modules may then be presented in the form of graphical and tabular summaries.

Module 1—Before HOV

Description

The Before HOV case (existing conditions) establishes the traffic volumes, level of service (LOS) and the impact values that form the basis for comparison with the After HOV conditions. The steps, conducted in a downward progression, are shown in Table 6-2.

Data

The input values are shown in the boxes of Table 6-2. These values comprise the current travel speed, vehicle volumes, modal split, vehicle occupancy, vehicle equivalences, and the length of the proposed HOV section.

Table 6-2 Module 1: Estimation of Existing Impacts

MODULE 1: BEFORE HOV, INPUTS AND IMPACTS

Item	Info. Source, or Calculation	HOVbus	HOVauto 2+	SOVs	TOTAL
Inputs: All 4 lanes (values per lane):					
Current (before HOV) travel speed, km/h	Surveys	24.00	24.00	24.00	
Vehicle volume by mode	Surveys, check consistency with speed	63	530	4992	5,585.00
Vehicle mode split, %	Mode % of total volume	1.13	9.49	89.38	100.00
Check total veh vol, pcphpl	Lookup la vol pcphpl × 4 … (4 × 1400)				5,600.00
Lane volume, pcphpl	HCM value divided by number of lanes				1,400.00
Ave. Vehicle Occupancy	Surveys	40.00	3.00	1.00	
Number of passengers	Vehicles × ave. occupancy	2,520.00	1,590.00	4,992.00	9,102.00
Pass. mode split, %	Mode % passengers of total	27.69	17.47	54.85	100.00
HOV section length, km.	Measured	8.00	8.00	8.00	8.00
Impacts:					
HOV veh running time, hours	HOV segment length/speed	0.33	0.33		
Emissions (CO), gm/veh km	Lookup	22.30	19.58	17.80	
Emiss/mode	Section length × vol. × emiss/veh km	11,239	83,019	710,861	805,119
Energy use, MJ/km	Lookup	4.45	1.82	1.65	
Energy use/mode, MJ	Length × vol. × MJ/veh km	2,243	7,696	65,894	75,833
Passenger travel time (hours)	Length × passengers/speed	840.00	530.00	1,664.00	3,034
Pass. modal split, % HOV users	All HOV pass/Total passengers, %	HOV passenger 45%			

All volumes are per hour unless stated otherwise

Unit values for emissions and energy use are derived from tabulations of vehicle fleet performance shown in the Appendices, and may be included but not necessarily illustrated in a worksheet as "lookup" values. The pollution and energy use data assume stable engine temperatures and energy use typical of vehicles that have been operating for at least 30 minutes. The impacts are based upon unit values for emissions (grams CO, for illustrative purposes), megajoules (MJ) for energy use, and passenger travel time as a measure of efficiency from the user's point of view, and to provide a basis for a preliminary cost and economic analysis.

Method and output

Each item of the analysis and the information sources or calculations are shown in the appropriate rows. The modes (HOVbus, HOVauto, and SOV) and the totals are shown in the columns. Note that the observed total traffic flow of the four lanes is 5,585 vph. This is close to the HCM (2) value of 5,600 (4 lanes at 1,400 vph each within LOS F, or interrupted, congested flow). This latter value, therefore, is used in the ensuing calculations to provide consistency with the traffic flow analyses in Module 2, which uses the tabulated speed-flow relationship of (2).

The method of estimating each value is shown in the second column of the table. Thus, under the existing conditions in the selected peak hour the number of passengers traversing the freeway is 9,102. The emissions are approximately 805,000 grams of CO, and the energy use is approximately 76,000 MJ. Total passenger time involved in traveling the 8-km freeway segment is approximately 3,000 hours, based upon a running time through the segment for all vehicles of 0.33 hr, or 20 minutes. Modal split expressed as a percentage of HOV passengers is 45%.

Module 2—After HOV

Description

This module is divided into three parts: demand estimation, traffic flow analysis, and impact estimation. The computations are summarized in Table 6-3, and are described in more detail in the following paragraphs.

Demand estimation

The procedure adopted in Chapter 5 is followed here. However, only a primary diversion and a modal shift are included in this example, and growth in traffic is not included. If growth were considered it would be included at this point by the appropriate factor. Both primary diversion and modal shift are

Table 6-3 Module 2: Estimated After HOV Traffic Flow and Impacts

MODULE 2: AFTER HOV TRAFFIC FLOW, DEMAND, AND IMPACTS-ADD-A-LANE

Item	Info. Source, or Calculation	LANE 1 (HOV)		LANES 2-5 (GP)			TOTAL
		HOVbus	HOVauto	Bus	HOVauto	SOVs	
Demand Estimate:							
Diversion factor	% of eligible bus and HOVauto to HOV lane	0.80	0.80				
Pass. vols. following diversion	Factor the "before HOV" pass volumes	2,016	1,272	504	318	4,992	9,102
Modal shift from SOVs	% of SOVs to bus and HOVautos					0.03	0
Pass. demand vols. from mode shift	Add shift to HOVs and deduct from SOVs	92	58			−150	0
Pass. volume by mode, lane	Add mode shift pass. to previous pass.	2,108	1,330	504	318	4,842	9,102
Vehicle volume by mode	Pass. volume/veh. occupancy	53	443	13	106	4,842	5,457
Traffic Flow Analysis:							
Gen demand vols. (per lane basis)	Bus+HOA in HOV lane, and ea. GP lane	HOV 496		GP 1,240			
Gen Purpose la. forced flow capacity	HCM value under 'before' conditions				1,400		
GP lanes, excess capacity, per lane	Demand vol. by lane-forced flow capacity				−160		
GP lanes excess capacity, total	Lane difference × number of lanes				−639		
Actual vols. by mode, per lane	GP lanes: split cap. vol. between modes	53	443	4	30	1,367	
After HOV speeds, km/h	Lookup (manual) HCM 94 and lane vol	90	90	24	24	24	
Impacts by Lanes:							
Impacts by Lane and Total							
HOV veh running time, hours	HOV segment length/speed	0.09	0.09				0.09
Vehicle volumes	Vehicle volumes by category	53	443	14	120	5,466	6,096
Passenger volumes	Mode vol × veh. occupancies	2,108	1,330	569	359	5,466	9,832
Emissions (CO), gm/veh km	Lookup	10.56	9.16	22.30	19.58	17.80	
Emiss/mode	Length × vol. × emiss/veh km	4,451	32,471	2,537	18,743	778,375	836,578
Energy use, MJ/km	Lookup	3.08	1.37	4.45	1.82	1.65	
Energy use/mode, MJ	Length × vol. × MJ/veh km	1,296	4,849	506	1,737	72,153	80,541
Passenger travel time (hours)	Length × passengers/speed	187	118	190	120	1822	2437
Pass. modal split, % HOV users	All HOV pass/Total passengers, %	HOV passenge 44%					

based upon passenger volumes and are later converted to vehicle volumes by dividing by the vehicle occupancy, in order to simplify the calculations. The passenger movements affected comprise the occupants of buses and HOVautos divided between the HOV lane and the GP lane, and SOVs in the GP lanes.

The values of the passenger diversions and the modal shift will be unique to each situation. For this example, the values used represent a low diversion and modal shift, based upon those described in Module 1. These values are as follows:

1. Primary diversion of existing HOV passengers on the freeway to the HOV lane: 80%.

2. Modal shift for HOV facilities: 3% of the freeway's SOV passengers to the HOV lane. This amount is based upon the estimated trip time savings (nearly 40% over an average total 15-mile commuting trip) related to modal shift, as shown in Figure 5-6. It is assumed further to be distributed to the existing HOV vehicles (buses and HOVautos) in the same proportion as these two modes currently exhibit.

The passenger and vehicle demands for the After HOV condition are computed by multiplying the volumes estimated in Module 1 by the primary diversion and latent demand specified above. This results in a redistribution of the passenger and vehicular traffic throughout the lanes. The number of passengers remains the same as for the Before HOV case, but the total number of vehicles is slightly less because of the modal shift of some passengers from SOVs to buses and HOVautos in the HOV lane. The volume of passengers will be modified as described under traffic flow analysis, below, because of the greater capacity of 5 lanes than 4.

Traffic flow analysis

This analysis deals primarily with the traffic flow in the GP lanes, because here the volume adjustments due to the diversion and modal shift will be apparent. As mentioned in Chapter 5, it is assumed that some vehicles from the hours preceding and following the peak hour will initially transfer to the GP lanes to replace the buses and HOVautos diverted and shifted from them. Therefore, each GP lane volume and speed is assumed to revert to approximately the Before HOV condition, and the total vehicle volume will increase by the HOV lane volume. The passenger volume will increase based upon the average vehicle occupancy of each vehicle category.

The transfer of passengers and vehicles between the hours of the peak period and the consequent effects on the volumes will almost certainly also change the overall travel times and costs between the relevant origins and destinations. This emphasizes the need for a comprehensive, regional, travel analysis to estimate long-term effects—an analysis that is beyond the scope of this book.

The final step of the traffic flow analysis is to insert the associated travel speed for the HOV lane vehicles into the calculations, based upon the HOV lane volumes in the HCM (2). The analyst performs the lookup manually in order to motivate a review for "reasonableness" before proceeding further. The speed for the HOV lane is 90 km/h. This is somewhat less than the free-flow speed of 97 km/h to allow for the predominant use of buses and possible lateral obstructions resulting from the HOV lane's physical features and geometry. The GP lanes, because there will be, at least initially, some added capacity, may operate at a slightly higher speed, but this has been ignored in these calculations. It is assumed that vehicles in the adjacent hours will fill any spare capacity, apart from "opening day" conditions, and the speed is shown to remain at 24 km/h. The passenger and vehicle volumes in the HOV lane and the GP lanes are shown in Figure 6-3, based upon the values calculated in this step of Module 2. The maximum congestion flow volume for the After HOV situation is shown as the same amount as for the Before HOV situation. Figure 6-3 provides a visual check on the results of the traffic flow analysis before calculating the impacts. Note that, because of the transfer of volumes to the peak hour, the adjacent lanes are more likely to experience higher speeds than during the peak hour.

Impact estimation

The key impacts resulting from the After HOV condition are listed as HOV vehicle (and passenger) running time, vehicle volume, passenger volume, CO emissions, energy use, and total passenger hours. This information, together with the Before HOV impacts, provides the basis for estimating the impact changes described in Module 3.

Module 3—Comparison of Impacts

As shown in Table 6-4, this module computes the algebraic sum of the impacts of the Before and the After HOV situations to give the changes in the impacts from Before HOV. Negative values indicate a reduction in the amount of the variable, and vice versa. As might be expected from the implementation of an add-a-lane HOV project, the major improvements are the decrease in the travel time of the HOV users and the decrease in the total travel time of all passengers. The latter occurs mainly because most of the bus passengers now travel in the HOV lane, at considerably higher speed. Total air pollution and energy use will increase, however, although some minimal decrease on a per person basis will occur.

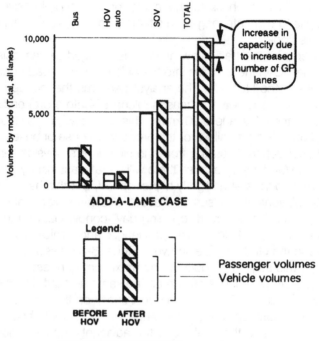

VOLUMES BY MODE

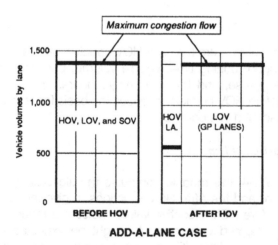

VOLUMES BY LANE

Figure 6-3 Comparison of mode and lane volumes

Table 6-4 Comparison of Before and After HOV Impacts

MODULE 3: DIFFERENCES BETWEEN
BEFORE HOV AND AFTER HOV IMPACTS

Impact item	Total change		Change per passenger	
	Amount	Percentage	Amount	Percentage
HOV vehicle running time, hr.	−0.24	−73	−0.24	−73
Vehicle volume, veh./hr.	496	9	0.0048	0.78
Passenger volume, Pass./hr.	730	8	NA	NA
CO emissions, gm.	31,459	4	−3.37	−4
Energy use, MJ	4,709	6	−0.14	−1.67
Passenger travel time, total hours	−597	−20	−0.09	−34
Pass. modal split, all HOV users (incl GP las.)	256	−1%	NA	NA

Modal split in terms of the passengers using transit and high occupancy automobiles is seen to decrease one percentage point, the shift to these modes being more than offset by the increased volume of SOVs in the GP lanes during the peak hour.

The results are also summarized as follows:

1. *HOV vehicle running time.* The greatest savings from the proposed facility will be the decrease in this running time. The nearly 15 minutes reduction is an approximately 73% savings over that of the Before HOV condition, and is the same on a per passenger basis.

2. *Vehicle volume.* The total vehicle volume has increased by nearly 500 vph, essentially the volume in the new HOV lane. This represents an approximately 9% increase over the Before HOV condition. On a per passenger basis, i.e., representing average vehicle occupancy, the change is less than 1%.

3. *Passenger volume.* Total passenger volume increases because of the added lane and in accordance with the vehicle occupancy associated with the increased vehicle volume. The passenger increase is 730, representing 8% of the Before HOV condition.

4. *CO air pollution emissions.* An increase of more than 31,000 grams, or 4%, occurs, based upon the calculated volumes and speeds. On a per passenger basis, the decrease is somewhat more than 2 grams per passenger, or 4%.

5. *Energy consumption.* An increase of over 4,700 MJ occurs, repre-senting nearly a 6% increase. On a per passenger basis, this amounts to a decrease of 3.4% from the Before HOV condition.

6. *Passenger travel time.* This decreases significantly; nearly 600 hours, or 20%, based upon the total number of hours. On a per passenger basis, the reduction is approximately 34%, due exten-sively to the larger proportion of passengers traveling at higher speeds in buses and HOVautos in the HOV lanes.

The above results are of particular interest when compared with those of the take-a-lane case, described in the next section.

Estimation of Impacts Using A Take-a-Lane HOV Project

This example has the same conditions as the previous one, except that the proposed project is to remain a total of four lanes, so that the After HOV condition is one HOV lane and 3 GP lanes.

The worksheet for the modules and the results are shown in Table 6-5. The inputs and calculations in Module 1 are the same as for the add-a-lane case. In Module 2, the use of three GP lanes results in a reduced total capacity in the GP lanes and, hence, a spillover volume of LOVs will occur. This spillover volume will tend to transfer to the hours immediately preceed-ing and following the peak hour. These events are shown graphically for the add-a-lane case and the take-a-lane case in Figure 6-4.

The differences in the impacts between the add-a-lane case and the take-a-lane case for the peak hour are clear from examination of Table 6-6 and Figure 6-5, where the results of each case are compared. Essentially, the HOV lane vehicle running time is the same, but all the other impacts are lower than in the take-a-lane case, in terms of the total amounts that occur in the peak hour. Effects on adjacent streets, particularly those parallel to the facility, are not discussed here, but could also be significant.

Table 6-5   Take-a-Lane HOV Analysis

MODULE 1: BEFORE HOV INPUTS AND IMPACTS

Item	Info. Source, or Calculation	HOVbus	HOVauto 2+	SOV	Total
Inputs: All 4 lanes (values per lane):					
Current (before HOV) travel speed, km/h	Surveys	24.00	24.00	24.00	
Vehicle volume by mode	Surveys, check consistency with speed	63	530	4992	5585.00
Vehicle mode split, %	Mode % of total volume	1.13	9.49	89.38	100.00
Check total veh vol, pcphpl	Lookup in vol pcphpl x 4... (4x1400)				5600.00
Lane volume, pcphpl	HCM value divided by number of lanes				1400.00
Ave. Vehicle Occupancy	Surveys	40.00	3.00	1.00	
Number of passengers	Vehicles x ave. occupancy	2520.00	1590.00	4992.00	9102.00
Pass. mode split, %	Mode % passengers of total	27.69	17.47	54.85	100.00
HOV section length, km.	Measured	8.00	8.00	8.00	8.00
Impacts:					
HOV veh running time, hours	HOV segment length / speed	0.33	0.33		
Emissions (CO), gm/veh km	Lookup	22.30	17.08	19.58	
Emiss/mode	Section length x vol. x emiss/veh km	11,239	75,472	781,947	868,658
Energy use, MJ/km	Lookup	4.45	1.82	1.65	
Energy use / mode, MJ	Length x vol. x MJ/veh km	2,243	7,696	65,894	75,833
Passenger travel time (hours)	Length x passengers / speed	840.00	530.00	1664.00	3,034
Pass. modal split, % HOV users	All HOV pass / Total passengers, %		HOV passengers 45%		

MODULE 2: AFTER HOV TRAFFIC FLOW, DEMAND, AND IMPACTS - TAKE-A-LANE
All volumes apcs per hour unless stated otherwise

Item	Info. Source, or Calculation	LANE 1 (HOV) HOVbus	LANE 1 (HOV) HOVauto	LANES 2-4 (GP) Bus	LANES 2-4 (GP) HOVauto	LANES 2-4 (GP) SOVs	TOTAL
Demand Estimate							
Diversion factor	% of eligible bus and HOVauto to HOV lane	0.80	0.80				
Pass. vols. following diversion	Factor the "before HOV" pass volumes	2,016	1,272	504	318	4,992	9,102
Modal shift from SOVs	% of SOVs to bus and HOVautos					0.03	
Pass. demand vols. from mode shift	Add shift to HOVs and deduct from SOVs	92	58			-150	0
Pass. volume by mode, lane		2,108	1,330	504	318	4,842	9,102
Veh. volume by mode	SOV Pass. / veh. occ.	53	443	13	106	4,842	5,457
Traffic Flow Analysis:							
Demand vols. (per lane basis)	Bus+HOA in HOV lane, and ea. GP lane	HOV 496		GP 1,654			
Gen Purpose la. forced flow capacity	HCM value under 'before' conditions				1,400		
GP lanes cap. deficiency, per lane	Demand vol. by lane-forced flow capacity				254		
GP lanes cap. deficiency, total	Lane difference x number of lanes				761		
Actual vols. by mode, per lane	GP lane: split cap. vol. between modes	53	443	4	30	1,367	
After HOV speeds	Lookup (manual) HCM 94 and lane vol	90	90	24	24	24	
Impacts by Lanes	Impacts by Lane and Total						
HOV veh running time, hours	HOV segment length / speed	0.09	0.09				
Vehicle volumes	Vehicle volumes by category	53	443	11	90	4,100	4,696
Passenger volumes	Mode vol x veh. occupancies	2,108	1,330	427	269	4,100	8,233
Emissions (CO), gm/veh km	Lookup	10.56	9.16	22.30	19.58	17.80	
Emiss/mode	Length x vol. x emiss/veh km	4,451	32,471	1,903	14,057	583,782	636,665
Energy use, MJ/km	Lookup	3.08	1.371	4.45	1.82	1.65	
Energy use / mode, MJ	Length x vol. x MJ/veh km	1,296	4,849	380	1,303	54,115	61,942
Passenger travel time (hours)	Length x passengers / speed	187	118	142	90	1367	1904
Pass. modal split, % HOV users	All HOV pass / Total passengers, %		HOV passengers 50%				

MODULE 3: DIFFERENCES BETWEEN BEFORE HOV AND AFTER HOV IMPACTS
All volumes apcs per hour values stated otherwise

IMPACT ITEM	Total Change Amount	Total Change Percentage	Change per Passenger Amount	Change per Passenger Percentage
HOV vehicle running time, hr.	-0.24	-73	-0.24	-73
Vehicle volume, veh./hr.	-904	-16	-0.04	-7
Passenger volume, Pass./hr.	-869	-10	NA	NA
CO emissions, gm.	-231,994	-27	-18	-19
Energy use, MJ	-13,891	-18	-1	-10
Passenger travel time, total hours	-1,130	-37	-0.10	-44
Pass. modal split, all HOV users (incl GP las.)	24	5%	NA	NA

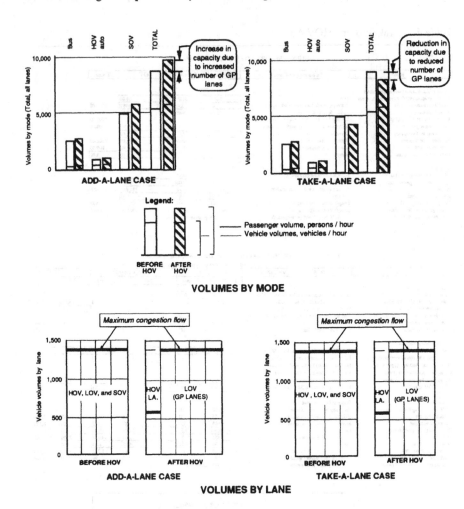

Figure 6-4 Comparison of peak hour mode and lane volumes for add-a-lane and take-a-lane cases

Table 6-6 Comparison of Add-a-Lane and Take-a-Lane Impacts

MODULE 3: DIFFERENCES BETWEEN
BEFORE HOV AND AFTER HOV IMPACTS—ADD-A-LANE

Impact item	Total change		Change per passenger	
	Amount	Percentage	Amount	Percentage
HOV vehicle running time, hr.	−0.24	−73	−0.24	−73
Vehicle volume, veh./hr.	496	9	0.0048	0.78
Passenger volume, Pass./hr.	730	8	NA	NA
CO emissions, gm.	31,459	4	−3.37	−4
Energy use, MJ	4,709	6	−0.14	−1.67
Passenger travel time, total hours	−597	−20	−0.09	−34
Pass. modal split, all HOV users (incl GP las.)	256	−1%	NA	NA

MODULE 3: DIFFERENCES BETWEEN BEFORE
HOV AND AFTER HOV IMPACTS—TAKE-A-LANE

Impact item	Total change		Change per passenger	
	Amount	Percentage	Amount	Percentage
HOV vehicle running time, hr.	−0.24	−73	−0.24	−73
Vehicle volume, veh./hr.	−904	−16	−0.04	−7
Passenger volume, Pass./hr.	−869	−10	NA	NA
CO emissions, gm.	−231,994	−27	−18	−19
Energy use, MJ	−13,891	−18	−1	−10
Passenger travel time, total hours	−1,130	−37	−0.10	−44
Pass. modal split, all HOV users (incl GP las.)	24	5%	NA	NA

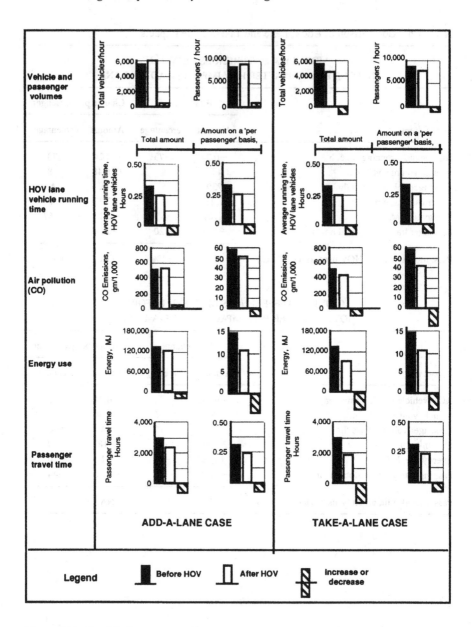

Figure 6-5 Graphical summary of impacts

Summarizing, the differences between the two cases and implications for the hours immediately preceding and following the peak hour are as follows:

Criteria	Add-a-lane case	Take-a-lane case
During the peak hour		
Passenger travel time		
HOV lane	Improved	Same as add-a-lane
GP lanes	Same or slight improvement, depending on spatial and temporal characteristics of congestion	No improvement—congestion encountered earlier in trip and for a longer period before and after the peak hour
Environmental impacts		
CO emissions		
Total amount	Increased	Decreased
Per passenger basis	Minimal reduction	Significant reduction
Energy use		
Total amount	Increased	Decreased
Per passenger basis	Minimal reduction	Significant reduction
Period hours immediately preceding and following the peak hour		
Passenger travel time		
HOV lane	No significant change from peak hr.	Same as add-a-lane
GP lanes	Probable significant improvement due to vehicles absorbed by increased capacity during peak hour	Probable worsening of congestion due to spillover from reduced GP lanes in peak hour
Environmental impacts		
CO emissions		
Total amount	Decreased due to reduced volume	Increased due to increased vols.
Per passenger basis	Minimal reduction	Significant reduction
Energy use		
Total amount	Decreased due to reduced volume	Increased due to increased vols.
Per passenger basis	Minimal reduction	Significant reduction

The take-a-lane case has more immediate benefits in terms of reduced environmental impacts. This is because of the reduced number of vehicles and passengers in the GP lanes and because the proportion of passengers (50% mode split versus 45%) using the HOV modes (which are both better environmentally and travel faster than the LOVs) is greater. During the hours preceding and following the peak hour, the total impacts will be greater for the take-a-lane case, and this may offset savings in adverse impacts during the peak hour. However, the average per passenger impacts will be less, again because of the greater proportional use of HOVs.

Concerning public acceptance, the add-a-lane case is likely to be more favorable with SOV users, because of the likelihood of their reduced travel time. Although this may be only minimal in the peak hour because of the assumed continued presence of a downstream bottleneck, the adjacent hours are likely to experience reductions in congestion because of the absorbtion of traffic into the peak hour. For the take-a-lane case the reduced adverse environmental effects throughout the peak period resulting from the greater proportional use of HOVs will assist in meeting air quality mandates and energy conservation efforts. However, the lower total capacity available from the loss of a GP lane and the consequent lengthening of queues and congestion may cause some dissatisfaction among SOV users, resulting in possible adverse perception of the project by segments of the public, and resulting difficulties in implementation.

Long-term effects are difficult to predict. As congestion has historically become worse, it is likely that the HOV facilities will encourage more people to use these modes. However, captive automobile users may perceive a reduction in accessibility and seek jobs in other locations. For both cases a spreading of the peak period and, possibly, some reduction in peak period volume because of a perceived reduction in the accessibility offered by the corridor for those not using the HOV lane are possible. However, the HOV lane will tend to enhance mobility for future users of HOVs, thus tending to emphasize the movement of people as opposed to vehicles. These concerns emphasize the need for more detailed, regional studies performed on a time series basis as planning progresses.

Review and Comments

The analysis method and process described in estimating the impacts of the proposed HOV facility in this chapter illustrate one approach among several. However, the approach shown includes the values of the impacts at each stage of the process to enable adjustments to be quickly made. If the worksheets are established in the form of computer spreadsheets, it is a relatively simple matter to adjust the input values and estimate the impacts on a "what if" basis. This may also be done, of course, by writing a computer program.

Although the method presented above is quite basic in approach, it may be refined in several ways. These include:

1. Different physical and operational features can be explored by changing the values and parameters associated with, for example, additional lanes

added to the freeway, either concurrent or contraflow, by modifying the format somewhat. Also, the use of 3+ or other car/vanpool occupancies eligible for using the HOV lanes can be explored.

2. The analysis process could be combined with a more detailed modal shift analysis procedure to provide initial estimates of HOV use and time savings to estimate a value of modal shift considered representative of actual conditions at the location being studied.

3. Input variables can be modified to reflect values of vehicle occupancy, volume/density/speed relationships, and a variety of scenarios related to diversion, latent demand (modal shift), and growth.

4. Unit values of impacts used for this analysis are basic and approximate. They do not take into account, for example, the effects of emissions and energy use with cold engine temperatures. Also, the effects of stops and starts are not included, and a factor can be incorporated in the analysis to account for these.

5. Values of diverted traffic, latent demand, and traffic growth can be determined for individual cases. Also, other factors those representing the extent of compliance with HOV regulations can be included.

The next section of this chapter provides a number of HOV projects to be solved. It will probably be advantageous for comparison and checking purposes to adopt the analysis techniques and formats described in the example provided earlier; however, this is not essential.

Projects for Solution

This section provides a range of input variables that enable the impacts of several alternative HOV proposals, similar to those in the preceding example, to be investigated. Estimation of these impacts will illustrate typical ranges of the impacts to be expected in practice. The work involved in solving each project should require about 4 to 6 hours of time if done manually and checked at each significant stage, and is therefore suitable as an individual project.

Scope of Work

The work to be done is to estimate the Before HOV impacts, the After HOV impacts, and the change in impacts for each of the 18 projects defined by the sets of parameters (Before HOV speed, primary diversion, and modal shift) listed as follows:

Project No.	Before HOV Speed (km/h)	Primary Diversion (%)	Modal Shift (%)	Project No.	Before HOV Speed (km/h)	Primary Diversion (%)	Modal Shift (%)
1	16	60	3	10	24	60	3
2	16	60	5	11	24	60	5
3	16	60	7	12	24	60	7
4	16	70	3	13	24	70	3
5	16	70	5	14	24	70	5
6	16	70	7	15	24	70	7
7	16	80	3	16	24	80	3
8	16	80	5	17	24	80	5
9	16	80	7	18	24	80	7

It is assumed that Before HOV speed reflects ongoing highway planning alternatives and that the effects of the diversion and mode shift levels are being explored to provide some estimate of the likely impacts. One or two projects may be undertaken by one person or in small groups.

For each of the 18 projects an "add-a-lane" and "take-a-lane" case should be examined and, for each of these, the impacts on a "total" and on a "per person" basis should be calculated, making a total of four sets of results for each project. This requirement is similar to that for the projects shown earlier, and it is suggested that the layouts and sequence of calculations be similar in order to facilitate comparisons and checks at key steps of the work. Adopting the same formats—or modified to suit different preferences—will assist in identifying obvious errors in the results, because the changes in costs and impacts should show consistency with the values of the Before HOV speeds, the primary diversions, and the modal shifts.

Analysis of combined results

As a combined group project the results of the individual projects may be plotted to provide an overview of the likely effectiveness of the HOV project and to make proposals for further alternatives and changes in the parameters that may assist in the decision-making process. A sample form for summarising the results of the projects is shown in Table 6-7. A separate form of this layout should be reproduced and labelled for each of the four sets of results mentioned above. Parameters may be changed by the analyst, for example, based upon less conservative estimates of diversion, and resulting from policy directions such as inducing a greater modal shift by decreasing park-and-ride costs for bus users, or by restricting the supply of downtown parking for commuters. To enable such comparisons to be made, plots of impacts related to diversion and to mode shift would be appropriate, and further relationships can be explored if required.

Table 6-7 Summary Sheet for Comparison of HOV Alternatives' Impacts

Alt. No.	Before HOV			After HOV			Change in Impacts		
	Energy (MJ)	CO Emiss. (Kg)	Running Time	Energy (MJ)	CO Emiss. (Kg)	Running Time	Energy (MJ)	CO Emiss. (Kg)	Running Time
			HOV Users (Hours) / All Users (Hours)			HOV Users (Hours) / All Users (Hours)			HOV Users (Hours) / All Users (Hours)
1			(Hours) (Hours)			(Hours) (Hours)			(Hours) (Hours)
2									
3									
4									
5									
6									
7									
8									
9									
10									
11									
12									
13									
14									
15									
16									
17									
18									

COMMENTS:

References

1. American Association of State Highway and Transportation Officials, *Guide for the Design of High-Occupancy Vehicle Facilities,* Washington, D.C., 1992.

2. Highway Research Board, *Highway Capacity Manual,* Special Report 209, third edition, National Research Council, Washington, D.C., 1994.

Appendices

Trip Rate Analysis

Shown on pages 176 and 177 are examples of trip rates for daily and peak periods for selected types of facilities.

Examples of Trip Rates

Source: National Cooperative Highway Research Program, Report No. 187, *Quick-Response Urban Travel Estimation Techniques and Transferable Parameters, Users Guide,* National Research Council, Washington, D.C., 1978

GENERATOR[b]	VEHICLE TRIPS[c] TO & FROM PER DAY PER		PERCENT TRIPS IN HOUR SHOWN			TYPICAL AUTO OCCUPANCY	TYPICAL % TRANSIT OF TOTAL PERSON TRIPS[d]
	DWELLING UNIT	ACRE	A.M. PEAK	P.M. PEAK	PEAK HR. OF GEN.		
Residential							
Single Family							
1 Du/acre	9.3	9.3	8.0	10.8	10.8	1.62	3.2
2 Du/acre	9.3	18.6	8.0	10.8	10.8	1.62	3.2
3 Du/acre	10.2	30.6	8.0	10.8	10.8	1.67	3.2
4 Du/acre	10.2	40.8	8.0	10.8	10.8	1.67	3.2
5 Du/acre	9.1	45.5	8.0	10.8	10.8	1.62	3.2
Medium Density							
(Duplex,							
Townhouses							
etc.)							
5 Du/acre	7.0	35.0	8.0	10.8	10.8	1.57	5.6
10 Du/acre	7.0	70.0	8.0	10.8	10.8	1.57	5.6
15 Du/acre	7.0	105.0	8.0	10.8	10.8	1.57	5.6
Apartments							
15 Du/acre	6.0	90.0	7.9	10.8	10.8	1.56	12.4
25 Du/acre	6.0	150.0	7.9	10.8	10.8	1.56	12.4
35 Du/acre	6.0	210.0	7.9	10.8	10.8	1.56	12.4
50 Du/acre	6.0	300.0	7.9	10.8	10.8	1.56	12.4
60 Du/acre	6.0	360.0	7.9	10.8	10.8	1.56	12.4
Mobile Home							
Park							
5 Du/acre	5.5	27.5	8.3	10.8	12.5	1.54	1.0
10 Du/acre	5.5	55.0	8.3	10.8	12.5	1.54	1.0
15 Du/acre	5.5	82.5	8.3	10.8	12.5	1.54	1.0
Retirement							
Community							
10 Du/acre	3.5	35.0	12.1	12.1	12.1	1.48	6.0
15 Du/acre	3.5	52.5	12.1	12.1	12.1	1.48	6.0
20 Du/acre	3.5	70.0	12.1	12.1	12.1	1.48	6.0
Condominiums							
10 Du/acre	5.9	59.0	7.1	7.1	7.1	1.56	9.0
20 Du/acre	5.9	118.0	7.1	7.1	7.1	1.56	9.0
30 Du/acre	5.9	177.0	7.1	7.1	7.1	1.56	9.c
Planned Unit							
Develop.							
5 Du/acre	7.9	39.5	10.1	10.1	10.1	1.58	7.1
15 Du/acre	7.9	118.5	10.1	10.1	10.1	1.58	7.1
25 Du/acre	7.9	197.5	10.1	10.1	10.1	1.58	7.1
	SEE INDIVIDUAL GENERATOR BELOW						
Miscellaneous							
Service	Station	Pump					
Station	748	133	1.5	3.0	4.0	1.55	-
Race Track	Seat	Attendee					
	0.61	1.08	-	-	-	2.05	-
Pro-Baseball	0.16	1.18	-	-	-	2.05	
Military Base	Military Personnel	Civilian Employees	Total Employees				
	2.2	7.1	1.8	-	-	-	1.42	-
	1000sq.ft. GFA	Employee	Acre				

Examples of Trip Rates (cont'd)

GENERATOR[b]	VEHICLE TRIPS[c] TO & FROM PER DAY PER			PERCENT TRIPS IN HOUR SHOWN			TYPICAL AUTO OCCUPANCY	TYPICAL % TRANSIT OF TOTAL PERSON TRIPS[d]
	1000sq.ft. GFA	EMPLOYEE	ACRE	A.M. PEAK	P.M. PEAK	PEAK HR OF GEN.		
Parks & Recreation (cont'd)								
National Monument	-	-	11.9	-	-	-	2.05	-
Ocean Front	-	-	21.6	-	-	-	2.05	-
Lake/Boating	-	-	3.6	-	-	-	2.05	-
Animal Attractions	-	-	72.2	-	-	-	2.05	-
Hospitals	Staff	Bed	Acre	-	-			
All Categories	6.1	14.8	40	-	-	11.7	1.40	-
General	5.9	14.0	-	18.0	9.0	-	1.42	17
Childrens	10.1	25.2	-	-	-	-	1.42	17
Convalescent	4.5	3.2	-	-	-	-	1.42	10
University	7.8	37.0	-	12.5	10.5	-	1.41	10
Veterans	2.2	3.8	-	11.0	16.5	-	1.32	10
Nursing Home	-	2.7	-	5.2	7.8	13.3	1.40	10
Clinics	5.9	-	-	-	-	-	1.40	10
Educational	Student	Staff						
All Categories	1.8	13.6		-	-	-	1.40	-
Four Year Univ	2.5	9.8		11.0	9.0	-	1.40	13
Jr.College	1.5	28.2		11.5	7.5	11.9	1.55	13
Secondary School	1.4	19.9		11.5	4.9	-	1.55	4
Elementary School	0.6	11.7		31.4	2.0	-	1.55	4 -e
Combined Elem/Sec.	0.8	11.8		-	-	-	1.55	4
Libraries	41.8	51.0		-	-	16.0	1.55	6
Airports	Take-Off/Landing	Employee	Acre					
General Aviation	2.5	6.5	3.6	11.8	10.5	15.7	1.52	1
Commercial	11.8	16.8	-	9.7	17.3	-	1.52	3
Hotel/Motel	Room	Employee						
Hotel	10.5	11.3		7.9	5.7	8.3	1.56	2
Motel	9.6	10.6		6.7	5.9	9.0	1.56	0
Resort Hotel	10.2	10.3		2.6	6.8	7.8	1.93	0

a. The trip rates given are based on a limited number of studies and thus must be used with caution. The ITE Trip Generation Report provides current data which is also periodically updated. The vehicle trip rates include external-internal and internal-external trip ends at generators as well as trucks, taxis and buses.

b. Most of the generators examined are located outside the central business districts of cities. The trip rates may thus be inapplicable to sites located within the dense urban core, particularly in large cities. Variations in generation rates may also exist because of the location of the generator either within a metropolitan area or outside that area.

c. The vehicle trip rates presented are actually volumes into and out of the site. As such, they may include some trips that would be passing the site on the adjacent street system, in any case, while making a trip for another reason, and they are induced to stop for impulse or convenience shopping, personal business or to drop off or pick up a passenger. The proportion of these trips has not been identified. Note also that ranges in trip rates can be expected and these can vary depending upon local conditions.

d. The typical transit % shown has a wide range of variation based on location within an urban area, level of service provided, etc., and as such, should be used only to provide gross approximations.

e. Does not include school bus transit.

Description of Freeway Levels of Service

The outline descriptions of freeway levels of service (LOS) described below augment the basic traffic speed-density-flow relationships outlined in Chapter 2. The source of the descriptions is the *Highway Capacity Manual*, Special Report No. 209, third edition, Transportation Research Board, Washington, D.C., 1994.

Operational characteristics for the six levels of service are shown in the illustrations that follow. The levels of service were defined to represent reasonable ranges in the three critical flow variables: speed, density, and service flow rate. General descriptions of operating conditions for each of the levels of service are as follows:

1. LOS A describes primarily free-flow operations. Average operating free-flow speeds generally prevail. Vehicles are almost completely unimpeded in their ability to maneuver within the traffic stream. Even at the maximum density for LOS A, the average spacing between vehicles is about 528 ft, or 26 car lengths, which affords the motorist a high level of physical and psychological comfort. Standing queues will generally not form, and traffic quickly returns to LOS A after passing any incident.

2. LOS B also represents reasonably free flow, and free-flow speeds are generally maintained. The lowest average spacing between vehicles is about 330 ft, or 18 car lengths. The ability to maneuver within the traffic stream is only slightly restricted, and the general level of physical and psychological comfort provided to drivers is still high.

3. LOS C provides for flow with speeds still at or near the free-flow speed. Freedom to maneuver within the traffic stream is noticeably restricted at LOS C, and lane changes require more vigilance on the part of the driver. Minimum average spacings are in the range of 220 ft, or 11 car lengths. The driver now experiences a noticeable increase in tension because of the additional vigilance required for safe operation.

4. LOS D is the level at which speeds begin to decline slightly with increasing flows. In this range, density begins to deteriorate somewhat more quickly with increasing flow. Freedom to maneuver within the traffic stream is more noticeably limited, and the driver experiences reduced physical and psychological comfort levels. Even minor incidents can be expected to create queuing, because the traffic stream has little space to absorb disruptions.

5. At its lower boundary, LOS E describes operations at capacity. Operations in this level are volatile, because there are virtually no usable gaps in the traffic stream. Vehicles are spaced at approximately six car

Illustration 3-5. LOS A.

Illustration 3-8. LOS D.

Illustration 3-6. LOS B.

Illustration 3-9. LOS E.

Illustration 3-7. LOS C.

Illustration 3-10. LOS F.

lengths, leaving little room to maneuver within the traffic stream at speeds that still exceed 50 mph. Maneuverability within the traffic stream is extremely limited, and the level of physical and psychological comfort afforded the driver is extremely poor.

6. The LOS F operations observed within a queue are the result of a breakdown or bottleneck at a downstream point. LOS F is used to describe conditions at the point of the breakdown or bottleneck as well as the operations within the queue that forms behind it.

    Whenever LOS F conditions exist, there is a potential for them to extend upstream for significant distances. A prerequisite for valid analyses using these procedures is the assumption that the section under consideration is free from downstream effects that propagate upstream. If upstream operations reflect the downstream bottleneck, they will not be as indicated by use of free-flow traffic relationships.

Capacities and Applicability of Transit: Selected Data

The data in this section provide some indication of the capacities of the various transit modes when selecting one or more modes for a corridor design. The approach described in Sections 3 and 4 will provide some guidance so the design does not provide excess capacity with a resulting inefficient use of resources.

Several options exist for allocating transit passengers between the various transit modes. Bus rapid transit (BRT), using the equivalent of a single freeway lane (considered separately from freeway operations but physically connected) for bus volumes of up to approximately 6,000 passengers per hour, is one method of accommodating high volumes of transit passengers. In terms of route capacity, the next highest option would be light rail transit (LRT) systems for passenger volumes predicted to be over 10,000 per hour on a single track. Another option may be to allocate all transit passengers to an LRT system, as long as the maximum volume is at least 6,000 passengers per hour. Therefore, transit passengers would be accommodated by the local bus system on the arterial streets mentioned above and the LRT system. Express bus or commuter bus services on the freeways in mixed traffic might then accommodate a relatively small number (say 2,000) passengers per hour. It is stressed that there is no single accepted method of conducting these analyses. Again, it is emphasized that the estimates of facility requirements being made are for a preliminary assessment only. For detailed planning an operational design studies, demand and system performance data of greater accuracy will be required. Detailed transit data is given on the following pages.

1. Transit modes related to residential density and rail mode threshold levels

Source: Gray, G.E., and L.A. Hoel, editors, *Public Transportation*, Prentice-Hall, Englewood Cliffs, NJ, 1992.

Transit Modes Related to Residential Density

Mode	Service	Minimum Necessary Residential Density (dwelling units/acre)	Remarks
Dial-a-bus	Many origins to many destinations	6	Only if labor costs are not more than twice those of taxis
	Fixed destinations or subscription service	3.5 to 5	Lower figure if labor costs are twice those of taxis; higher if thrice those of taxis
Local bus	Minimum, ½-mi route spacing, 20 buses/day	4	
	Intermediate, ½-mi route spacing, 40 buses/day	7	Average, varies as a function of downtown size and distance from residential area to downtown
	Frequent, ½-mi route spacing, 120 buses/day	15	
Express bus reached on foot	5 buses during 2-h peak period	15 (average density over 2 mi^2 tributary area)	From 10 to 15 mi away to largest down towns only
reached by auto	5 to 10 buses during 2-h peak period	3 (average density over 20 mi^2 tributary area)	From 10 to 20 mi away to downtowns larger than 20 million ft^2 of non-residential floor space
Light rail	5-min headways or better during peak hour	9 (average density for a corridor of 25 to 100 mi^2)	To downtown of 20 to 50 million ft^2 of non-residential floor space
Rapid transit	5-min headways or better during peak hour	12 (average density for a corridor of 100 to 150 mi^2)	To downtown of larger than 50 million ft^2 of nonresidential floor space
Commuter rail	20 trains a day	1 to 2	Only to largest down-towns, if rail line exists

Source: Boris S. Pushkarev and Jeffrey M. Zupan, *Public Transportation and Land Use Policy*, a Regional Plan Association Book (Bloomington, Ind.: Indiana University Press, 1977).

Threshold Volumes for Rapid Transit Development
(keyed to type of structure)
(minimum service frequency, 8 min)

Mode	Type of Construction	Daily Pass.-Mi/Mi of Route
Rail rapid	Above ground	14,000
	One-third tunnel	17,000–24,000[a]
	All tunnel	24,000–42,000[a]
LRT	Low capital	4000
	Considerable grade separation	7000
	One-fifth in tunnel	13,500
	All tunnel	40,000
Downtown people mover	Above ground	12,000
	All tunnel	30,000

[a]Range reflects varying criteria for cost/weekday passenger-mi of travel

Source: Adapted from Boris Pushkarev, with Jeffrey M. Zupan and Robert S. Cumella, *Urban Rail in America: An Exploration of Criteria for Fixed-Guideway Transit* (Bloomington, Ind.: Indiana University Press, 1982), p. 116.

2. Capacities of various modes

Source: Vuchic, Vukan R., *Urban Public Transportation, Systems and Technology,* Prentice-Hall, Englewood Cliffs, NJ, 1981.

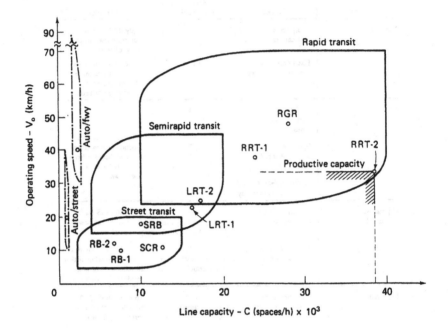

		Generic Class	Private		Street Transit		Semirapid Transit		Rapid Transit	
Characteristic	Mode Unit		Auto on street	Auto on freeway	RB	SCR	SRB	LRT	RRT	RGR
1. Vehicle capacity, C_v	sps/veh		4–6 total, 1.2–2.0 usable		40–120	100–180	40–120	110–250	140–280	140–210
2. Vehicles/transit unit	veh/TU		1	1	1	1–3	1	1–4	1–10	1–10
3. Transit unit capacity	sps/TU		4–6 total, 1.2–2 0 usable		40–120	100–300	40–120	110–600[b]	140–2000	140–1800
4. Maximum technical speed, V	km/h		40–80	80–90	40–80	60–70	70–90	60–100	80–100	80–130
5. Maximum frequency, f_{max}	TU/h		600–800	1500–2000	60–120	60–120	60–90	40–90	20–40	10–30
6. Line capacity, C	sps/h		720–1050[b,d]	1800–2600[d]	2400–8000	4000–15,000	4000–8000	6000–20,000	10,000–40,000	8000–35,000
7. Normal operating speed, V_o	km/h		20–50	60–90	15–25	12–20	20–40	20–45	25–60	40–70
8. Operating speed at capacity, V_o^c	km/h		10–30	20–60	6–15	5–13	15–30	15–40	24–55	38–65
9. Productive capacity, P_c	(sp-km/ h[2]) × 10[3]		10–25[b]	50–120	20–90	30–150	75–200	120–600	400–1800	500–2000
10. Lane width (one-way)	m		3.00–3.65	3.65–3.75	3.00–3.65	3 00–3 50	3.65–3.75	3.40–3.75	3.70–4.30	4 00–4 75
11. Vehicle control[c]	—		Man./vis.	Man /vis	Man./vis.	Man./vis.	Man./vis.	Man./vis.-sig.	Man.-aut./sig.	Man -aut./sig.
12. Reliability	—		Low–med.	Med.–high	Low–med	Low–med	Low–med	High	Very high	Very high
13. Safety	—		Low	Low–med.	Med.	Med.	High	High	Very high	Very high
14. Station spacing	m		—	—	200–500	250–500	350–800	350–800	500–2000	1200–4500
15. Investment cost per pair of lanes	($/km) × 10[6]		0.2–2.0	2 0–15 0	0 1–0 4	1 0–2 0	3.0–9 0	3.5–12.0	8 0–25.0	10.0–25.0

[a] Abbreviations: sps = spaces; veh = vehicles; TU = transit unit
[b] Values for C and P_c are not necessarily products of the extreme values of their components, because these seldom coincide.
[c] For auto, lane capacity, for transit, line (station) capacity in TU/h
[d] For private auto capacity is product of average occupancy (1 2–1 3) and f_{max} since all spaces cannot be utilized
[e] Abbreviations are for: manual, visual, signal, and automatic

3. Peak hour bus volumes

Source: Transportation Research Board, National Academy of Sciences, *Highway Capacity Manual,* Special Report 209, TRB, Washington, D.C., 1985.

FACILITY OR SOURCE	BUSES PER HOUR	HEADWAY (SEC)	AVERAGE BUS STOP SPACING (FT)	AVERAGE BUS SPEED (MPH)	PASS. PER HOUR	REMARKS
Freeway or Busway						
Lincoln Tunnel Uninterrupted Flow	735	4.9	No Stops	30	32,560	Connects to Midtown bus terminal
I-495 (New Jersey) Exclusive Bus Lane, Uninterrupted Flow	485	7.3	No Stops	30–40	21,600	
San Francisco Oakland Bay Bridge	350	10.3	No Stops	30–40	13,000	Pre-BART connects to Transbay terminal
Shirley Highway Busway, Wash., D.C.	200	18.0	No Stops	35(Freeway)	10,000	900-ft stop spacing in CBD
Bus-Only Mall						
State Street, Chicago	180	20.0	400	0–5	9,000	Based on peak 15-min rate
Portland, 5th at 6th Ave.	180	20.0	NA	5–10	9,000	
Arterial Street						
Michigan Ave., Chicago	228	15.0	NA	NA	11,400	Some multiple lane use, 5-min rate
Madison Ave., N.Y.C.	200 ±	18.0	1,000	NA	10,000	Two exclusive bus lanes
Hillside Ave., N.Y.C.	170	21.0	530	Not Cited	8,500[a]	Multiple lane use with lightly patronized stops
14th Street, Wash., D.C.	160	23.0	900	5–12	8,000	Approach to CBD
Market St., Philadelphia	150	24.0	300–600	5–10	6,100–9,900	Multiple lanes— Pre-Chestnut St. mall
K Street, Wash., D.C.	130	28.0	500	5–8	6,500	Pre-Metro
Main St., Rochester	80	45.0	1,000	5	4,000	Some platooning at stops
Downtown Streets with Stops (Various Cities)	80–120	30.0–45.0	500	5–10	4,500–6,000[a]	

[a] Estimated, assuming 50 passengers per bus; (1 ft = 0.305 m; 1 mph = 1.6 kph)

SOURCE: Compiled from various bus-use studies—1972-1978 conditions. Summarized in Ref. *34.*

4. Light rail transit (LRT) volumes

Source: U.S. Department of Transportation, Urban Mass Transportation Administration, *Characteristics of Urban Transportation Systems,* Washington, D.C., 1985.

SERVICE VOLUME OF TYPICAL LIGHT RAIL TRANSIT SYSTEMS (PEAK HOUR)

Location	Vehicles Per Hour	Headway[1] (seconds)	Actual Passenger Loads	Average Trip Length (Miles)
Cologne	59	61	9,600	3.2
Rotterdam	37	97	4,600	n/a
Dusseldorf	92	39	14,000	2.9
Frankfurt	23	157	8,200	2.7
Stuttgart	40	90	1,200	3.5
Hanover	80	45	18,000	3.4
Gothenburg	88	41	7,200	2.7
Bielefeld	24	150	4,300	2.5

[1] Numbers are based on a single one-way track; as service volume increases, special signals are necessary.

Source: Vuchic, V., Light Rail Transit Systems - A Definition and Evaluation, U.S. Department of Transportation, October 1972.

5. Rail rapid transit (RRT) volumes

Source: U.S. Department of Transportation, Urban Mass Transportation Administration, *Characteristics of Urban Transportation Systems,* Washington, D.C., 1985.

SERVICE VOLUME OF TYPICAL RAIL RAPID TRANSIT LINES (PEAK HOURS)

Location-Facility	Trains Per Hour	Headway (seconds)	Cars Per Train	Cars Per Hour	Seating Capacity Per Car	Seating Capacity Per Train	Total	Actual Passenger Loads
New York-IND-6th-8th-Ave. Express	32	112	10	320	60	600	19,200	61,400
New York-IRT-Lexington Ave. Express	31	116	9	279	40	360	11,160	44,510
New York-IND - 8th Ave. Express	30	120	10	300	60	600	18,000	62,030
New York-IRT - 7th Ave. Express	24	150	9	216	40	360	8,640	36,770
Toronto-Yonge St. Subway	28	128	8	224	62	496	13,888	35,166
Chicago Congress St. Expressway	25	144	6	150	49	294	7,350	10,376
Cleveland-Rapid Transit Line	20	180	6	120	53	318	6,360	6,211
Philadelphia PATCO	30	120	6	180	80	480	14,400	36,000
San Francisco BART[1]/	6	600	10	60	72	720	4,320	12,720
Boston MBTA - Red Line	15	240	4	60	64	256	3,840	14,340
Chicago Dan Ryan Line	30	120	8	240	50	400	12,000	24,000

[1]/Headways were improved after opening of Transbay Tunnel.

Sources: Institute of Traffic Engineers, Capacity and Limitations of Urban Transportation Modes, Washington, D.C., 1965.

Transportation Systems Center, Safety and Automatic Train Control for Rail Rapid Transit System, U.S. Department of Transportation, July 1974.

6. Right-of-way requirements

Source: Vuchic, Vukan R., *Urban Public Transportation, Systems and Technology,* Prentice-Hall, Englewood Cliffs, NJ, 1981.

Mode	Schematic of R/W	Line capacity reserve	Terminal area requirements
Private autos on street (Persons/vehicle: 1.3 Maximum freq.: 700)	|◄── 17 Lanes x 3.50m ──►| ⌐□□□ ┘ |◄── 119m ──►|	None	Parking: 23 m² /person For 15,000 people 34.5 ha (85 acres)
Private autos on freeway (1.3: 1800)	7 Lanes x 3.65m |◄──►| □□□ ⌐ ┐ |◄── 51m ──►|	None	Same as above, plus interchanges
Regular buses (R/W C) (75; 100)	4 Lanes x 3.50m ⌐□□ ┘ ◄─| 14m |◄──	None (station and way capacities reached)	Each station 20 x 80 m on the surface
Semirapid buses (artic., R/W B) (100; 90)	2 Lanes x 3.65m + shoulders □ ◄─| 11m |◄	None (station capacity reached, way capacity not)	Each station 25 x 100 m on the surface
Light rail transit (2 artic. car trains) (400; 50)	2 tracks □ ⫠ ⟋⟋⟋⟋ or: or: ▭ ◄─►|7.5 m|◄ ─	33%	Each station from 12 x 50m on the surface to 20 x 90 grade separated
Rail rapid transit (1000; 25 RGR, 1000; 40 RRT)	2 tracks ⟋⟋⟋⟋⟋ □ ⟋⟋⟋⟋ ◄─| 8m |◄ ─ ⟋‾‾⟍	67–167%	Each station from 20 x 100 to 25 x 210m grade separated. No surface occupancy

Unit Values for Impact Analysis

The values listed in the table that follows are those used in the worksheet modules of Chapters 5 and 6. The values are based upon the following sources:

1. Traffic volumes are based upon Transportation Research Board, National Academy of Sciences, *Highway Capacity Manual,* Special Report 209, third edition, TRB, Washington, D.C., 1994.

2. Energy use and air pollution emission rates are based upon U.S. Department of Transportation, Urban Mass Transportation Administration, *Characteristics of Urban Transportation Systems,* Washington, D.C., 1992.

Speed km/h	Volume pcphpl	Air pollution, gm/veh km			Energy Use, mJ/veh km		
		AUTO	IOV AUT	BUS	AUTO	HOV AUTO	BUS
16	950	30	33	31	1.9	2.09	5.5
17	1010	27.75	30.53	29.63	1.8688	2.055625	5.36875
18	1070	25.5	28.05	28.25	1.8375	2.02125	5.2375
19	1230	23.25	25.58	26.88	1.8063	1.986875	5.10625
20	1200	21	23.1	25.5	1.775	1.9525	4.975
21	1250	20.2	22.22	24.7	1.7438	1.918125	4.84375
22	1300	19.4	21.34	23.9	1.7125	1.88375	4.7125
23	1350	18.6	20.46	23.1	1.6813	1.849375	4.58125
24	1400	17.8	19.58	22.3	1.65	1.815	4.45
25	1431	17	18.7	21.5	1.6188	1.780625	4.31875
26	1462	16.2	17.82	20.7	1.5875	1.74625	4.1875
27	1493	15 4	16.94	19.9	1.5563	1.711875	4.05625
28	1524	14.6	16.06	19.1	1.525	1.6775	3.925
29	1555	13.8	15.18	18.3	1.4938	1.643125	3.79375
30	1586	13	14.3	17.5	1.4625	1.60875	3.6625
31	1617	12.65	13.92	17.15	1.4313	1.574375	3.53125
32	1648	12 3	13.53	16.8	1.4	1.54	3.4
33	1679	11.95	13.15	16.45	1.3813	1.519375	3.36875
34	1710	11.6	12.76	16.1	1.3625	1.49875	3.3375
35	1741	11.25	12.38	15.75	1.3438	1.478125	3.30625
36	1772	10.9	11.99	15.4	1.325	1.4575	3.275
37	1803	10.55	11.61	15.05	1.3063	1.436875	3.24375
38	1835	10.2	11.22	14.7	1.2875	1 41625	3.2125
39	1869	9.85	10.84	14.35	1.2688	1.395625	3.18125
40	1900	9.5	10 45	14	1.25	1.375	3.15
41	1919	9.15	10.07	13.65	1.2313	1.354375	3.11875
42	1938	8.8	9.68	13.3	1.2125	1.33375	3.0875
43	1956	8.45	9.295	12.95	1.1938	1.313125	3.05625
44	1975	8.1	8.91	12.6	1.175	1.2925	3.025
45	1994	7.75	8.525	12.25	1.1563	1.271875	2.99375
46	2013	7 4	8.14	11.9	1.1375	1.25125	2.9625
47	2031	7.05	7.755	11.55	1.1188	1.230625	2.93125
48	2050	6.7	7.37	11.2	1.1	1.21	2.9
49	2069	6.35	6.985	10.85	1.1102	1.221224	2.9125
50	2088	6	6.6	10.5	1.1204	1.232449	2.925
51	2106	5.95	6.545	10.43	1.1306	1.243673	2.9375
52	2125	5.9	6.49	10.37	1.1408	1.254898	2.95
53	2144	5.85	6.435	10.3	1.151	1.266122	2.9625
54	2163	5.8	6.38	10.23	1.1612	1.277347	2.975
55	2181	5.75	6.325	10.17	1 1714	1.288571	2.9875
56	2200	5.7	6.27	10 1	1.1816	1.299796	3
57	2204	5 65	6.215	10.03	1.1918	1.31102	3.0125
58	2208	5.6	6.16	9.967	1.202	1.322245	3.025
59	2213	5.55	6 105	9.9	1.2122	1.333469	3.0375
60	2217	5.5	6.05	9.833	1.2224	1.344694	3.05
61	2221	5.45	5.995	9.767	1.2327	1.355918	3.0625
62	2225	5 4	5.94	9.7	1.2429	1.367143	3.075
63	2229	5.35	5.885	9.633	1.2531	1.378367	3.0875
64	2233	5.3	5.83	9.567	1.2633	1.389592	3 1
65	2238	5.25	5.775	9.5	1.2735	1.400816	3.1125
66	2242	5.2	5.72	9.433	1.2837	1.412041	3 125
67	2246	5.15	5.665	9.367	1.2939	1.423265	3.1375
68	2250	5.1	5.61	9.3	1.3041	1.43449	3.15
69	2254	5.05	5.555	9.233	1.3143	1.445714	3.1625
70	2258	5	5.5	9.167	1.3245	1.456939	3.175
71	2263	4.95	5.445	9.1	1.3347	1.468163	3.1875
72	2267	4.9	5.39	9.033	1.3449	1.479388	3.2
73	2271	4.85	5.335	8.967	1.3551	1.490612	3.252
74	2275	4.8	5.28	8.9	1.3653	1.501837	3.304
75	2279	4.75	5.225	8.833	1.3755	1.513061	3.356
76	2283	4.7	5.17	8.767	1.3857	1 524286	3.408
77	2288	4 65	5.115	8.7	1.3959	1.53551	3.46
78	2292	4.6	5.06	8.633	1.4061	1.546735	3.512
79	2296	4.55	5.005	8.567	1.4163	1.557959	3.564
80	2300	4.5	4.95	8.5	1.4265	1.569184	3.616
81	2289	4.88235	5.371	8.706	1.4367	1.580408	3.668
82	2278	5.26471	5.791	8.912	1.4469	1.591633	3.72
83	2267	5.64706	6.212	9.118	1.4571	1.602857	3.772
84	2256	6.02941	6.632	9.324	1.4673	1.614082	3.824
85	2245	6.41176	7.053	9.529	1.4776	1.625306	3.876
86	2233	6.79412	7.474	9.735	1.4878	1.636531	3.928
87	2222	7.17647	7.894	9.941	1.498	1.647755	3.98
88	2211	7.55882	8.315	10.15	1.5082	1.65898	4.032
89	2200	7.94118	8.735	10.35	1.5184	1.670204	4.084
90	2125	8.32353	9.156	10.56	1.5286	1.681429	4.136
91	2050	8.70588	9.576	10.76	1.5388	1.692653	4 188
92	1975	9.08824	9.997	10.97	1.549	1.703878	4.24
93	1900	9.47059	10.42	11.18	1.5592	1.715102	4.292
94	1825	9.85294	10.84	11.38	1.5694	1.726327	4.344
95	1750	10.2353	11.26	11.59	1.5796	1.737551	4 396
96	1675	10.6176	11.68	11.79	1.5898	1 748776	4.448
97	1600	11	12.1	12	1 6	1.76	4.5

Conversion Factors Between Si and English Units

Source: Excerpts from Davis, Stacy C., *Transportation Energy Data Book: Edition 14*, Oak Ridge National Laboratory, U.S. Department of Energy, Oak Ridge, TN, 1995.

SI Prefixes and Their Values

	Value	Prefix	Symbol
One million million millionth	10^{-18}	atto	a
One thousand million millionth	10^{-15}	femto	f
One million millionth	10^{-12}	pico	p
One thousand millionth	10^{-9}	nano	n
One millionth	10^{-6}	micro	µ
One thousandth	10^{-3}	milli	m
One hundredth	10^{-2}	centi	c
One tenth	10^{-1}	deci	
One	10^{0}		
Ten	10^{1}	deca	
One hundred	10^{2}	hecto	
One thousand	10^{3}	kilo	k
One million	10^{6}	mega	M
One billion[a]	10^{9}	giga	G
One trillion[a]	10^{12}	tera	T
One quadrillion[a]	10^{15}	peta	P
One quintillion[a]	10^{18}	exa	E

[a]Care should be exercised in the use of this nomenclature, especially in foreign correspondence, as it is either unknown or carries a different value in other countries. A "billion," for example signifies a value of 10^{12} in most other countries.

Metric Units and Abbreviations

Quantity	Unit name	Symbol
Energy	joule	J
Specific energy	joule/kilogram	J/kg
Specific energy consumption	joule/kilogram·kilometer	J/(kg·km)
Energy consumption	joule/kilometer	J/km
Energy economy	kilometer/kilojoule	km/J
Power	kilowatt	Kw
Specific power	watt/kilogram	W/kg
Power density	watt/meter^3	W/m^3
Speed	kilometer/hour	km/h
Acceleration	meter/second^2	m/s^2
Range (distance)	kilometer	km
Weight	kilogram	kg
Torque	newton·meter	N·m
Volume	meter^3	m^3
Mass; payload	kilogram	kg
Length; width	meter	m
Brake specific fuel consumption	kilogram/joule	kg/J
Fuel economy (heat engine)	liters/100 km	L/100 km

Energy Unit Conversions

1 Btu = 778.2 ft-lb	1 kWhr = 3412 Btu[a]
= 107.6 kg-m	= 2.655×10^6 ft-lb
= 1055 J	= 3.671×10^5 kg-m
= 39.30×10^{-5} hp-h	= 3.60×10^6 J
= 39.85×10^{-5} metric hp-h	= 1.341 hp-h
= 29.31×10^{-5} kWh	= 1.360 metric hp-h
1 kg-m = 92.95×10^{-4} Btu	1 J = 94.78×10^{-5} Btu
= 7.233 ft-lb	= 0.7376 ft-lb
= 9.806 J	= 0.1020 kg-m
= 36.53×10^{-7} hp-h	= 37.25×10^{-8} hp-h
= 37.04×10^{-7} metric hp-h	= 37.77×10^{-8} metric hp-h
= 27.24×10^{-7} kWh	= 27.78×10^{-8} kWh
1 hp-h = 2544 Btu	1 metric hp-h = 2510 Btu
= 1.98×10^6 ft-lb	= 1.953×10^6 ft-lb
= 2.738×10^6 kgm	= 27.00×10^4 kg-m
= 2.685×10^6 J	= 2.648×10^6 J
= 1.014 metric hp-h	= 0.9863 hp-h
= 0.7475 kWh	= 0.7355 kWh

[a] Electricity generation and distribution efficiency is approximately 29%.

1 in. = 83.33×10^{-3} ft	1 ft = 12.0 in.
= 27.78×10^{-3} yd	= 0.333 yd
= 15.78×10^{-6} mile	= 189.4×10^{-3} mile
= 25.40×10^{-3} m	= 0.3048 m
= 0.2540×10^{-6} km	= 0.3048×10^{-3} km
1 mile = 63360 in.	1 km = 39370 in.
= 5280 ft	= 3281 ft
= 1760 yd	= 1093.6 yd
= 1609 m	= 0.6214 mile
= 1.609 km	= 1000 m

1 ft/sec = 0.3048 m/s = 0.6818 mph = 1.0972 km/h

1 m/sec = 3.281 ft/s = 2.237 mph = 3.600 km/h

1 km/h = 0.9114 ft/s = 0.2778 m/s = 0.6214 mph

1 mph = 1.467 ft/s = 0.4469 m/s = 1.609 km/h

Bibliography

AASHTO. A *Policy on Geometric Design of Highways and Streets-1990*.

American Association of State Highway and Transportation Officials, Washington, D.C., 1990.

American Public Transit Association. *Light Rail Transit. Washington, D.C.,* September 1987.

Barton-Aschman Associates. *Traveler Response to Transportation System Changes.* Report DOT-FH-II-9579. Federal Highway Administration, Washington, D.C., July 1981.

Batchelder, J.H., M. Golenberg, J.A. Howard, and H.S. Levinson. *Simplified Procedures for Evaluating Low Cost TSM Projects.* NCHRP Report No. 263. Transportation Research Board, Washington, D.C., October 1983.

Batz, T. *High Occupancy Vehicle Treatments, Impacts, and Synthesis.* Report No. FHWA/NJ-86-017-7767. Federal Highway Administration, Washington, D.C., 1986.

Box, P.C., and J.C. Oppenlander. *Manual of Traffic Engineering Studies,* 4th ed. Institute of Transportation Engineers, Washington, D.C., 1976.

Cervero, R. *Suburban Gridlock.* Center for Urban Policy Research, Rutgers University, New Brunswick, NJ, 1986.

Chang, M.F., et al. *The Influence of Vehicle Characteristics, Driver Behavior, and Ambient Temperature on Gasoline Consumption in Urban Areas.* General Motors Corporation, Warren, MI, 1976.

Charles River Associates, Inc., and Herbert S. Levinson. *Characteristics of Urban Transportation Demand—An Update,* rev. ed. Prepared for UMTA, Technology Sharing Program. Report No. DOT-T-88-18. U.S. Department of Transportation, Washington, D.C., July 1988.

Cheslow, M.D. *Potential Use of Carpooling During Periods of Energy Shortages; Considerations in Transportation Energy Contingency Planning.* Special Report 191. Transportation Research Board, National Research Council, Washington, D.C., 1980, pp. 38-43.

———. *Proceedings of the Conference on Energy Contingency Planning in Urban Areas.* Special Report 203. National Research Council, Washington, D.C., 1983.

Curry, D.A., and D.G. Anderson. *Procedures for Estimating Highway User Costs, Air Pollution, and Noise Effects.* National Cooperative Highway Research Program Report 133. Transportation Research Board, National Research Council, Washington, D.C., 1972.

Davis, Stacy C. *Transportation Energy Data Book: Edition 14.* Oak Ridge National Laboratory, U.S. Department of Energy, Oak Ridge, TN, 1995.

Department of Civil Engineering and Urban Planning. *Planning and Designing a Transit Center-Based Transit System: Guidelines and Examples from Case Studies in 22 Cities.* Department of Civil Engineering and Urban Planning, University of Washington, Seattle. Prepared for USDOT and UMTA, September 1980, DOT-I-81-5.

Dickey, John W., Senior author, *Metropolitan Transportation Planning,* 2d ed. Hemisphere Publishing Corporation, New York, 1983.

Eno Foundation for Transportation. *Transportation in America: A Statistical Analysis of Transportation in the United States,* 8th ed. EFT, Washington, D.C., 1990.

Environmental Protection Agency. *User's Guide to MOBILE 1: Mobile Source Emissions Model.* Office of Air, Noise, and Radiation, EPA, Washington, D.C., 1978.

Federal Highway Administration. *Guidelines for Trip Generation Analysis.* U.S. Department of Transportation, U.S. Government Printing Office, Washington, D.C., June 1967.

————. *Trip Generation Analysis.* U.S. Department of Transportation, U.S. Government Printing Office, Stock No. 050-001-00101-2, Washington, D.C., 1975.

————. *Computer Programs for Urban Transportation Planning: PLANPAC/BACKPAC General Information.* U.S. Department of Transportation, U.S. Government Printing Office, Stock No. 050-001-00125-0, Washington, D.C., April 1977.

————. *Traveler Response to Transportation System Changes,* 2d ed. U.S. Department of Transportation, Washington, D.C., July 1981.

————. *Transportation Management for Corridors and Activity Centers: Opportunities and Experiences.* Washington, D.C., May 1986.

————. *Quantification of Urban Freeway Congestion and Analysis of Remedial Measures.* Report FHWA/RD-87/052. Washington, D.C., October 1986.

Fertal, M.J., et al. *Modal Split: Documentation of Nine Methods for Estimating Transit Usage.* Bureau of Public Roads, U.S. Department of Commerce, U.S. Government Printing Office, Washington, D.C., December 1966.

Fielding, G.J. *Managing Public Transportation Strategically: A Comprehensive Approach to Strengthening Service and Monitoring Performance.* Jossey-Bass Publishers, San Francisco, CA, 1987.

Gray, G.E., and L.A. Hoel, eds. *Public Transportation,* 2d ed. Prentice-Hall, Englewood Cliffs, NJ, 1992.

Highway Research Board. *Parking Principles.* Highway Research Special Report 125. National Academy of Sciences, Washington, D.C., 1971.

Hanson, Susan. *The Social Geography of Urban Transportation.* The Guilford Press, New York, 1986.

Horowitz, J.L. *Air Quality Analysis for Urban Transportation Planning.* MIT Press, Cambridge, MA, 1982.

Hutchinson, B.G. *Principles of Urban Transport Systems Planning.* McGraw-Hill, New York, 1974.

Institute of Transportation Engineers. *Transportation and Traffic Engineering Handbook,* 2d ed. ITE, Washington, D.C., 1982.

————. "Strategies to Alleviate Traffic Congestion." *Proceedings of ITE's 1987 National Conference.* San Diego, CA, March 8-11, 1987. ITE, Washington, D.C., 1987.

————. *The Effectiveness of High Occupancy Vehicles Lane.* ITE, Washington, D.C., 1988.

————. *Transportation and Traffic Engineering Handbook.* Prentice-Hall, Englewood Cliffs, NJ, 1989.

Jacobs, M., R.E. Skinner, and A.C. Lemer. *Transit Project Planning Guidance: Estimation of Transit Supply Parameters.* Prepared by the Transportation Systems Center for UMTA. Washington, D.C. Urban Mass Transportation Administration, UMTA-MA-09-9015-85-01, October 1984.

Khisty, C. Jotin. *Transportation Engineering-An Introduction.* Prentice-Hall, Englewood Cliffs, NJ, 1990.

Lang, A.S., and R.M. Soberman. *Urban Rail Transit: Its Economics and Technology.* MIT Press, Cambridge, MA, 1964.

Larsen, R.I. *Air Pollution from Motor Vehicles.* The New York Academy of Science, New York, 1966.

May, A.D. *Traffic Flow Fundamentals.* Prentice-Hall, Englewood Cliffs, NJ, 1990.

Mayworm, P., A.M. Lago, and J.M. McEnroe. *Patronage Impacts of Changes in Transit Fares and Services.* Prepared for USDOT, Office of Methods and Demonstration. Report No. 1205-UT. US DOT, Washington, D.C., 1980.

McShane, W.R., and R.P. Roess. *Traffic Engineering.* Prentice-Hall, Englewood Cliffs, NJ, 1990.

Organisation for Economic Co-operation and Development. *Traffic Measurement Methods for Urban and Suburban Areas.* OECD Publications, Paris, 1979.

Papacostas, C.S., and P.D. Prevedouros. *Transportation Engineering and Planning.* Prentice-Hall, Englewood Cliffs, NJ, 1993.

Perkins, H.C. *Air Pollution.* McGraw-Hill, New York, 1974.

Public Technology Inc. *Manual on Planning and Implementing Priority Techniques for High-Occupancy Vehicles.* Report DOT-OS-60076. Federal Highway Administration, Washington, D.C., June 1977.

Pushkarev, B., and J. Zupan. *Public Transportation and Land-Use Policy.* Indiana University Press, Bloomington, IN, 1977.

————. *Urban Rail in America: An Exploration of Criteria for Fixed Guideway Transit.* Regional Plan Association. U.S. DOT Report No. UMTA-NY-06-0061080-1, November 1980.

Raus, J. *A Method for Estimating Fuel Consumption and Vehicle Emissions on Urban Arterials and Networks*. Report FHWA-TS-81-210. Office of Research and Development, Federal Highway Administration, Washington, D.C., 1981.

Scheel, J.W. *A Method for Estimating and Graphically Comparing the Amounts of Air Pollution Emissions Attributable to Automobiles, Buses, Commuter Trains, and Rail Transit*. Automotive Engineering Congress, Detroit, MI, 1972.

Sosslau, A.B., A.B. Hassam, M.M. Carter, and G.V. Wickstrom. *Quick-Response Urban Travel Estimation Techniques and Transferable Parameters, User's Guide*. National Cooperative Highway Research Program Report 187. National Research Council, Washington, D.C., 1978.

Spielberg, F., et al. *Evaluation of Freeway High-Occupancy Vehicle Lanes and Ramp Metering*. Report DOT-OST078-050. Federal Highway Administration, Washington, D.C., March 1980.

Transportation Research Board. "Transportation and Air Quality." *TR News* 148, Special Issue, 1990.

―――. *Highway Capacity Manual*. Special Report No. 209, 3d ed. TRB, Washington, D.C., 1994.

The Urban Analysis Group. *TRANSPLAN User's Manual*. UAG, Danville, CA, 1990.

Urban Mass Transportation Administration. *Guidelines for Preparing Environmental Assessments*. Circular # UMTA C 5620.1, Washington, D.C., October 16, 1979.

―――. *Draft Alternatives Analysis Procedures and Technical Guidelines: Appendix A, Estimating of Transit Supply Parameters*. UMTA, 1980.

Urbitran Associates. *Transportation Systems Management: Implementation and Impacts*. Urban Mass Transportation Administration, Washington, D.C., March 1982.

U.S. Department of Transportation. *Organization and Content of Environmental Assessment Materials*. Notebook 5. U.S. Government Printing Office, Stock No. 050-000-00109-1, Washington, D.C., 1975.

―――. *Characteristics of Urban Transportation Systems*. Report No. UMTA-IT-06-0049-79-1, Washington, D.C., 1979.

Vuchic, Vukan, R. *Urban Public Transportation, Systems and Technology*. Prentice-Hall, Englewood Cliffs, NJ, 1981.

Wayson, R.L. *Transportation Planning and Air Quality*. American Society of Civil Engineers, New York, 1992.

Wilbur Smith Associates. *Transportation and Parking for Tomorrow's Cities*. New Haven, CT, 1966.

Wright, Paul H., and Norman S. Ashford. *Transportation Engineering-Planning and Design* 3d ed. John Wiley and Sons, New York, 1989.

Yu, J.C. *Transportation Engineering*. Elsevier North-Holland, New York, 1982.

Zerbe, R.O., and K. Croke. *Urban Transportaion for the Environment*. Ballinger Publishing, New York, 1974.

Index

Activity systems,
   relation between transportation and, 32
   urban, 31, 32, 96
Air quality, (*see* Impacts)
Air quality improvement program, 5
Ambient air quality standards, 6
Assignment, trip, (*see* Demand)
Attractions, 51, 99,100, (*see also* Demand, trip
   distribution)
Automobile
   capacity and level-of-service, 66–68, 112
   impacts, 14–19, 112–120
      air pollution, 86–89, 118–120
      costs, 79–83, 114–116
      energy use, 84–87, 117–118
      in modal mix, 69–74
      in project, 112–122
   occupancy, 9, 93, 95, (*see also* HOV facilities,
      eligibility)
   ownership, in trip generation, 40, 41, 100
Average daily traffic (ADT), 67, 68,

Bus
   capacity and level of service, 69–71, 112
   fleet size, 69, 116
   impacts, 14–19, 112–120
      air pollution, 86–89, 118–120
      costs, 79–83, 114–116
      energy use, 84–87, 117–118
   in modal mix, 69–75
   occupancy, 95, 111, 112
   speed, in fleet size estimates, (see Fleet size)

Calibration of models, 36

Capacity and level of service
   automobiles (highway), 66–68, 95, 112
   bus system, 69–71, 95, 112
   light rail system, 69–71, 95, 112
   rail rapid transit, 69–71, 95, 112
Carpal, 18, (*see also* HOV facilities)
Category analysis, 39–44, 100, (*see also* Cross-
   classification analysis, and
Demand, trip generation)
City, transportation project layout, 93
Congestion, (*see also* HOV facilities)
   effects of increases in, 8
   management system (CMS), 7
   mitigation, 5,
Corridor
   circumferential, 10, 11
   in project, 92
   radial, 10, 11
   transportation, 21–26
   volumes, peak hour vehicle, 23, 109
Costs, (*see* Impacts, for automobile, bus, light
   rail transit, and rail rapid transit)
Criteria, related to values, goals, objectives, and
   standards, 3, 4
Cross-classification analysis, 39–44, 100(*see
   also* Category analysis, and Demand, trip
   generation)
Cross-sections
   for highway and transit facilities, 66
   for HOV facilities, 134,155

Data sources
   census tracts, 33
   zones, 33

195